Praise for the hardcover edition

"In this lucid road map for the nascent discipline of 'de-extinction,' Shapiro, an evolutionary biologist, examines not only how we can resurrect long-vanished species but also when we cannot or should not."—*Scientific American*

"As Shapiro sees it, de-extinction isn't about geeky genetic sleight of hand or about the resurrection of legendary beasts; it's a valuable new tool for conserving and enriching the global ecosystem."—*Natural History*

"[A] disturbing and thoughtful new book....Shapiro makes a good, sensible, balanced case."—Cathy Gere, *The Nation*

"[A] thoughtful roadmap....Readers will emerge with the ability to think more deeply about the facts of de-extinction and cloning at a time when hyperbolic and emotionally manipulative claims about such scientific breakthroughs are all too common." —*Publishers Weekly*

"Some of the best conversations I've had in recent months have come about while discussing de-extinction. The concept is simple: should we clone extinct animals, *Jurassic Park*-style, from found genetic material? How do we do it? What would the impact be on the environment? Shapiro makes it clear that we should have this discussion now because the future of de-extinction is real and coming fast."—Andrew Sturgeon, *Flavorwire*, from "10 Must-Read Academic Books of 2015"

"[T]houghtful and well-written....Shapiro does an excellent job of showing that the realities of genuine science can be as exciting as the fantasies of science fiction." —Nick Rennison, *Daily Mail*

"[A] fascinating book....A great popular science title, and one that makes it clear that a future you may have imagined is already underway."—*Library Journal*, starred review

"In *How to Clone a Mammoth*, molecular paleontologist Beth Shapiro spells out, step by step, how and how soon real scientists might be able to bring an extinct species back to life."—Nancy Szokan, *Washington Post*

"Shapiro is an acute, lively, sceptical and nuanced writer"—Caspar Henderson, *Spectator*

"Shapiro's thought-provoking book offers excitement and wonder—but also comes with a warning. We must think carefully, not just about how we can achieve this incredible scientific feat, but also about where it is likely to have the most positive (or least negative) impact, and why it is worth the investment and associated risks. [...] She paints a scientifically accurate yet magical world where Pleistocene giants might roam the Arctic tundra once again, and where we have the chance to undo some past mistakes—as long as we remember to keep looking towards the future."—Tiffany Taylor, *Times Higher Education*

"Beth Shapiro's 'how-to' manual couldn't be more timely."—*New Scientist*, a *New Scientist* best reads from 2015 selection

"Shapiro's book is a thoughtful how-to guide for the painstaking process of reviving not just mammoths but passenger pigeons and other lost species. Her aim is to separate science from science fiction by taking a critical look at proposals for bringing these animals back."—Allison, Bohac, *Science News*

"From her front-row seat as one of the pioneers of ancient-DNA research, Shapiro explains the fieldwork, lab science, and prospective ecology involved with the so-far hypothetical endeavor."—Bob Grant, *The Scientist*

"*How To Clone A Mammoth* is about as close as you get to sitting down with a nice cup of tea to have a decent chinwag with a mate about resurrecting the woolly mammoth.... Refreshingly, she replaces hyperbole with humour to guide the reader through the basics of de-extinction science... that personal touch brings warmth."—Dr Tori Herridge, *BBC Focus Magazine*

"Skilfully combining accounts of the scientific problems with ethical and practical considerations, the book is an informative and at times highly entertaining account of the life of a modern mammoth hunter."—William Hartston, *Daily Express*

"[Shapiro] goes to great lengths to demystify the art and science of cloning."—*Kirkus Reviews*

"Shapiro's book is fascinating."—*The Irish Examiner*

"This book is an excellent introduction to the emergent science of de-extinction."—*Choice*

"From basic science to ethics, *How to Clone a Mammoth* is a thorough and captivating exploration of an area at the leading edge of conservation biology. This book educates readers and entices all of us to delve more deeply into the issues discussed."—Simon Levin, author of *Fragile Dominion*

"Beth Shapiro is an evolutionary biologist who specialises in ancient DNA....Who better to take us through the technological developments and evidentiary likelihood of recreating extinct species? [A] well-written factual summary...playfully set out."—David Callahan, *Birdwatch*

"Shapiro has done an excellent job."—Ian Simmons, *Fortean Times*

"In *How to Clone a Mammoth*, Shapiro provides detailed descriptions of current state-of-the-art bioengineering technologies, explaining just what can and cannot be done. Readers of the book will be well equipped to develop their own informed opinions on this controversial topic."—Ravi Mandalia, Techie News

"In her new book, Shapiro offers an accessible, rigorous, I-can't-believe-it's-not-sci-fi guide to the world of de-extinction research. You can read the book as a pop primer on genetics, a field guide to future fauna, or as a roadmap to the next generation of conservation science. But reading about these mammoths and Tasmanian tigers, you start to feel that Shapiro is getting at bigger questions....[She] is a lucid, relaxed, and often hilarious guide to the strange world of people who try to resurrect dead species."—Michael Schulson, *Religion Dispatches*

"[Shapiro] has skillfully blended cutting edge science with an overview of the ramifications that resurrecting lost fauna might have for the restoration of declining ecosystems."—Everything Dinosaur Blog

"[W]arm and accessible...Shapiro's informal approach, peppered with deadpan asides, is a welcome change from the hyperbole and grandstanding that have come to characterise popular debates on rewilding and de-extinction....The open-hearted simplicity of *How to Clone a Mammoth* provides a great entry point for people who want to join in [the conversation]."—Tori Herridge, *Literary Review*

"[Shapiro] goes to great lengths to demystify the art and science of cloning."—*Kirkus Reviews*

"[Beth Shapiro's] book exposes the fallacies in our thinking about such activities, as well as the real possibilities and even potential values of restoring some extinct species. This is not a silly book; rather, it is a serious story well told and a fun read."—*Buffalo News*

HOW TO CLONE A MAMMOTH

HOW TO CLONE A MAMMOTH

THE SCIENCE OF DE-EXTINCTION

BETH SHAPIRO

PRINCETON UNIVERSITY PRESS
PRINCETON AND OXFORD

Published by Princeton University Press, 41 William Street,
Princeton, New Jersey 08540
In the United Kingdom: Princeton University Press,
6 Oxford Street, Woodstock, Oxfordshire OX20 1TR

press.princeton.edu

Third printing, and first paperback printing, 2016

Cloth ISBN 978-0-691-15705-4

Paper ISBN: 978-0-691-17311-5

The Library of Congress has cataloged the cloth edition of this book as follows:
Shapiro, Beth Alison.
How to clone a mammoth : the science of de-extinction / Beth Shapiro.
p. cm
Includes bibliographical references and index.
ISBN 978-0-691-15705-4 (hardcover : alk. paper) 1. Extinct animals—Genetics. 2. Extinct animals—Cloning. 3. DNA, Fossil. 4. Extinction (Biology) I. Title.
QL88.S49 2015
591.68—dc23
2014049574

British Library Cataloging-in-Publication Data is available

This book has been composed in Baskerville Original Pro and Trade Gothic

Printed on acid-free paper. ∞

Printed in the United States of America

3 5 7 9 10 8 6 4

For my children, James and Henry, who will inherit whatever mess we make.

CONTENTS

PROLOGUE

The first use of the word "de-extinction" was, as far as I can tell, in science fiction. In his 1979 book *The Source of Magic*,[1] Piers Anthony describes a magician who suddenly finds himself in the presence of cats, which, until that moment, he had believed to be an extinct species. Anthony writes, "[The magician] just stood there and stared at this abrupt de-extinction, unable to formulate a durable opinion." I imagine that this is precisely how many of us might react to our first encounter with a living version of something we thought was extinct.

The idea that de-extinction might actually be possible—that science might advance eventually to the point where extinction is no longer forever—is both exhilarating and terrifying. How would de-extinction change the way we live? Would de-extinction provide new opportunities for economic growth and galvanize global conservation efforts? Or would it lull us into a false sense of security and ultimately increase the rate of species extinction?

The year 2013 saw "de-extinction" become its own new branch of science, at least according to the *Times*.[2] Despite this lofty status, there is as yet no consensus as to what the goal of de-extinction science is. At first, it seems obvious. De-extinction aims to resurrect, via cloning, identical copies of extinct species. For species that have been extinct for a long time, however—the passenger pigeon, the dodo, the mammoth—cloning is not a viable option. In the case of these species, de-extinction will have to mean something else. Most likely, it will mean that specific traits and behaviors of the extinct species will be genetically engineered into living species. The living species would then gain the adaptations necessary to thrive where the extinct species once did. Will society, however, respond favorably to de-extinction if the goal is not to bring back an actual mammoth, dodo, or passenger pigeon?

Piers Anthony's novel was eerily prescient with regard to our reaction to de-extinction. Immediately after his magician accepted that de-extinction was possible and, presumably, in the midst of forming an opinion about it, Anthony's magician had another thought. Anthony writes, "If [the magician] killed these animals, would he be re-extincting* the species?"

*A warning to grammarians: In this sentence, Anthony's magician provides what is perhaps the earliest example of that awkward moment when, while discussing some aspect of de-extinction, it suddenly becomes clear that there is no satisfying way to complete a thought without offending people like you. How should one designate a species that has been brought back to life? While "de-extinction" seems perfectly logical with reference to a process, to refer to the end-result of that process as "de-extinct" seems inappropriate. "De-extincted," while more logical, is painful to write down, much less to say out loud. I prefer "unextinct" to "de-extinct," as "unextinct" seems to describe the state of being, rather than the process of getting there. One might say, for example, "The mammoth is unextinct." Of course, "The mammoth is no longer extinct" is certainly sufficient.

What, then, is the present participle? I shudder when I say that George Church's lab is de-extincting the mammoth. My reaction to the phrase, however, has nothing to do with the science. At the first formal scientific discussion of de-extinction, some of us suggested using "resurrection" and its various conjugates, as in, "We are resurrecting extinct species." While "resurrect" makes perfect grammatical sense, however, its religious connotations seemed misleading. Certainly, "De-extincting," is terrible, as is "re-extincting." Yet, here we are. While it takes longer to say it, perhaps we should stick with something that is simple to understand and entirely inoffensive—at least grammatically: We are developing the science necessary to bring extinct species back to life.

Many of the people with whom I interact believe that de-extinction is inevitable. I'll admit, however, that this is a biased sample of the population, and that most people are likely to care about de-extinction only insofar as it might affect them personally. Some people of course love the idea of de-extinction. They may be swayed and enthused by the idea that resurrected species might improve wild habitats. Or they may just want the opportunity to see and touch a mammoth. Other people, including very sensible and intelligent people, hate the idea of de-extinction, citing both the high cost of resurrecting extinct species and the myriad risks of reintroducing organisms into the wild whose environmental impacts are—because they are extinct—necessarily unknown. Those people who fear de-extinction the most, like Anthony's magician, take solace in its reversibility. This worries me. It is undoubtedly true that history repeats itself and that, if need be, we could *re*-eradicate any species we brought back. However, our goal as scientists working in this field is not to create monsters or to induce ecological catastrophe but to restore interactions between species and preserve biodiversity. If we do arrive at a time when science makes it possible to resurrect the past, it might take years or decades to see the results of this work. I certainly hope we do not simply turn around at the first signs of imperfection and destroy what we worked so hard to accomplish.

Certainly, if we are to make room for extinct species—or for hybrids of extinct and living species—in the real world, we as a society will have to alter our attitudes, our actions, and even our laws. Science is paving the way to resurrect the past. The road, however, will be long, not necessarily direct, and certainly not smooth.

With this book, I aim to provide a road map for de-extinction, beginning with how we might make the decision about what species or trait to resurrect, traveling through the circuitous and often confusing path from DNA sequence to living organism, and ending with a discussion about how to manage populations of engineered individuals once they are released into the wild. My goal is to explain de-extinction in a way that separates science from science fiction. Some steps in the de-extinction process, such as finding well-preserved remains of extinct species, will be relatively simple to complete. Others, however, such as cloning extinct species, may never be feasible. My perspective, as a scientist who is actively involved in de-extinction research, is that of an enthusiastic realist. I believe that de-extinction is in many cases scientifically and ethically unjustified. However, I also believe that de-extinction technology has great potential to become an important tool for conserving species and habitats that are threatened in the present day. If that seems paradoxical, read on.

HOW TO CLONE A MAMMOTH

CHAPTER 1

REVERSING EXTINCTION

A few years ago, a colleague of mine practically bit my head off for getting the end date of the Cretaceous period wrong by *a little bit*. I was presenting an informal seminar about my research to graduate students at my university, which at the time was Penn State. My seminar was about mammoths—in particular, about when, where, and why mammoths went extinct, or at least what we've learned about the mammoth extinction by extracting bits of mammoth DNA from frozen mammoth bones. Before talking about this very recent extinction, I opened with a discussion of older and more famous extinctions. My offending slide cited the date for the end of the Cretaceous period and beginning of the Paleogene, also known as the K-Pg boundary and best known as the time of the extinction of the dinosaurs, at "around 65 million years ago." That date, I was told, was inexcusably imprecise. The K-Pg boundary occurred 65.5 ± 0.3 million years ago (at least that was the scientific consensus of the time), and I was *not* to be forgiven those 200,000 to 800,000 years.

While I appreciate that my fellow academics would have preferred meticulous attention to detail, I did not bring up the dinosaurs to discuss the precise timing of their demise. My goal was simply to make the point that while we think we now know why the dinosaurs went extinct so many millions of years ago, we still argue about what caused extinctions that took place within the last ten thousand years. Did the mammoths and other ice age animals go extinct because Earth's climate was suddenly too warm to support them? Or did our ancestors hunt them to death? The question remains open, perhaps because we are not particularly comfortable with the answer.

The last dinosaurs went extinct after a massive asteroid struck just off the coast of Mexico's Yucatan Peninsula. Similar cataclysmic events—major explosive volcanic eruptions or impacts of large asteroids or comets—are thought to have caused the other four mass extinctions in Earth's history. Each time, dense clouds of dust and other debris were suddenly ejected into the atmosphere, blocking out the sunlight. Without sunlight, the plants suffered and many species died. As the plant communities collapsed, so did the animals that ate the plants, and then the animals that ate the animals that ate the plants, and so on up the food chain until somewhere between 50 percent and 90 percent of all species that were alive at the time of the catastrophic event became extinct.

The mammoth extinction is different. We know of no single catastrophic event that happened within the last 10,000 years that might have caused mammoths to go extinct. Recent genetic research shows that mammoth populations probably started to decline sometime during or just after the peak of the last ice age some 20,000 years ago, as the rich arctic grasslands—often called the steppe tundra—on which they relied for food were gradually replaced by modern arctic vegetation. Mammoths were extinct in continental North America and Asia by around 8,000 years ago but survived for another few thousand years in two isolated locations in the Bering Strait: the Pribilof Islands off the western coast of Alaska, where mammoths survived until around 5,000 years ago, and Wrangel Island off the northeastern coast of Siberia, where they survived until around 3,700 years ago.

We know from the fossil record that mammoths, steppe bison, and wild horses dominated the Arctic landscape for a long time before the peak of the last ice age. In fact, they were the most abundant large mammals in the North American Arctic for most of the last 100,000 years. This was a very cold period of Earth's history and included two ice ages—one that peaked at around

80,000 years ago and another that peaked around 20,000 years ago—separated by a long cold interval. It was only after the peak of the most recent ice age that the climate really began to warm up, transitioning into the present warm interval (the Holocene epoch) by around 12,000 years ago. Because mammoths, steppe bison, and wild horses disappeared only *after* the Holocene had begun, it is reasonable to conclude that these species may simply have been adapted to living in a cold climate. When the world warmed up, the cold-adapted went extinct.

While this explanation is attractively simple, it has some problems. Most importantly, while we know from the fossil record that woolly mammoths lived in North America throughout at least the last 200,000 years, that period does not include only very cold intervals. In fact, around 125,000 years ago, Earth was as warm as or warmer than it is today. This was the peak of what we call the last interglacial period, which lasted from around 130,000 years ago until the beginning of the ice age around 80,000 years ago. Remains of mammoths, steppe bison, and wild horses are found in the fossil record of the last interglacial, indicating that they were able to survive despite the warmer climate. Their bones were, however, much less abundant during the interglacial than they were during the later, cold interval. According to the fossil record from the interglacial, a different community of animals dominated the warm Arctic from that which dominated when it was cold. The community of the interglacial period included giant sloths, camels, mastodons, and giant beavers: animals that were adapted to life in a warm climate.

If we look further back in time in the fossil record, a pattern begins to emerge. The Pleistocene epoch lasted from around 2.5 million years ago until around 12,000 years ago, when the Holocene epoch began. During the Pleistocene, our planet experienced at least twenty major shifts between cold, glacial intervals (ice ages) and warmer interglacial intervals. Average temperatures swung a whopping 5°–7°C with each climatic shift. Glaciers advanced or retreated, causing plants and animals to scramble (figuratively) to find suitable habitat. When the climate was cold, cold-adapted species were widespread. When it was warm, these cold-adapted species survived in isolated patches of refugial habitat, often at the edges of their former ranges. During the warm periods, warm-adapted species were widespread, and these warm-adapted species became restricted to warm refugia when it was cold. Range shifts were common during the Pleistocene, but extinctions were rare. And then, around 12,000 years ago, the climate swung from cold to warm, just as it had many times before. This time, however, cold-adapted fauna did not simply become less abundant. This time, many of them went extinct.

What was different about this most recent climate shift? The answer is not entirely clear. However, one potential explanation stands out: By the beginning of the Holocene, a new species had appeared on nearly every continent. This new species had a remarkably big brain and a capacity to transform its habitat to suit its needs, rather than seek habitats to which it was best adapted. This species was also alarmingly destructive. Wherever it went, its arrival seemed to coincide with the extinction of other, mostly large-bodied species. This species was, of course, humans.

Was it our fault that mammoths and other ice age animals went extinct? Interestingly, there is strong evidence that climate, and not humans, may have triggered the declines toward extinction. Humans and mammoths lived together in the arctic regions of Europe and Asia for many thousands of years during the last of the Pleistocene ice ages. The archaeological record shows that humans did hunt mammoths during this time, but since mammoths survived until much later, this hunting pressure was clearly not sufficient to drive mammoths to extinction. In North America, there is even clearer evidence that climate is to blame for diminishing mammoth populations. Humans did not arrive in North America until well after the populations of mammoths, steppe bison, and wild horses had already begun to decline toward extinction. Given this evidence, it is tempting to conclude that these extinctions were not our fault. After all, if we were not there, we could not have done it.

It is important, however, to understand the difference between declining populations and disappearing populations. Estimates of population size based on the fossil record or from genetic data can pinpoint when species began to decline from their ice age peaks but not when they actually went extinct. If we focus on disappearance rather than on decline, it is difficult to say with confidence that humans did not play a pivotal role in these extinctions. Populations of cold-adapted animals declined during every warm interval, not just during the most recent warm interval. In the past, however, these populations survived by finding and hiding out in refugial habitats, biding their time until the next cold period got under way. They probably did exactly that when the present warm interval began. This behavior, however, may have made them more vulnerable to extinction once humans were in the picture.

Ultimately, mammoths, steppe bison, and wild horses probably went extinct because of a combination of climate change, human hunting, and the disappearance of the steppe tundra. Rapid warming after the last ice age led to a decline in crucial habitat. Fewer herbivores trampling and consuming the vegetation meant that nutrients recycled more slowly, reducing the productivity of the ecosystem. To make matters worse, a new and intelligent predator appeared that was capable of zeroing in on any remaining ice-age habitat as ideal hunting grounds. Growing human populations and increasingly sophisticated human technologies further isolated these refugial populations from each other and from the resources they needed to survive. For some species, refugial populations may have held on well past the beginning of the Holocene. For example, our DNA work has shown that steppe bison survived in isolated patches in the far northern Rocky Mountains until as recently as one thousand years ago. As we learn more about the timing and pattern of these and other recent extinctions, there is little doubt that the role of humans will become increasingly clear.

THE SIXTH EXTINCTION

More than 3,700 years after the last mammoth died on Wrangel Island, we are witnessing an alarming number of contemporary extinctions, and the rate of extinction appears to be increasing. Some scientists have gone so far as to refer to the Holocene extinctions as the Sixth Extinction, suggesting that the crisis in the present day has the potential to be as destructive to Earth's biodiversity as the other five mass extinctions in our planet's history.

The word alone—extinction—frightens and intimidates us. But why should it? Extinction is part of life. It is the natural consequence of speciation and evolution. Species arise and then compete with each other for space and resources. Those that win survive. Those that lose go extinct. More than 99 percent of species that have ever lived are now extinct. Indeed, our own species' dominance is possible only because the extinction of the dinosaurs made space for mammals to diversify, and eventually we outcompeted the Neandertals.

I think people are scared of extinction for three reasons. First, we fear missed opportunities. A species that is lost is gone forever. What if that species harbored a cure for some terrible disease or was critically important in keeping our oceans clean? Once that species is gone, so is that opportunity. Second, we fear change. Extinction changes the world around us in ways that we both can and cannot anticipate. Every generation thinks of our version of the world as the authentic version of the world. Extinction makes it harder for us to recognize and feel grounded in the world we know. Third, we fear failure. We enjoy living in a rich and diverse world and feel an obligation, as the most powerful species that has ever lived on this planet, to protect this diversity from our own destructive tendencies. Yet we chop down forests and destroy habitats. We hunt and poach species even when we know they are perilously close to extinction. We build cities, highways, and dams and block migration routes between populations. We pollute the oceans, rivers, land, and air. We move around as fast as we can on airplanes, trains, and boats and introduce foreign species into previously undisturbed habitats. We fail to live up to our obligation to protect or even coexist with the other species with which we share this planet. And when we stop to think about it, it makes us feel terrible.

Extinction is much easier for us to swallow when it is clearly *not* our fault. Why did the mammoth go extinct? As humans, we want the answer to be *something natural*. Natural climate change, for example. We would prefer to learn that mammoths went extinct because they needed the grasslands of the steppe tundra to survive and that they simply starved to death as the steppe tundra disappeared after the last ice age. We would prefer not to learn that mammoths went extinct because our ancestors greedily harvested them for their meat, skins, and fur.

While some of us may not care about extinction as long as we are not personally affected, many of us find extinction unacceptable, particularly if it is our fault. Most contemporary extinctions are easy to ignore, as they have little influence on our day-to-day lives. The cumulative effect of these extinctions is, however, a future of very reduced biodiversity. This future could be one in which so many changes have occurred to the terrestrial and marine ecosystems that we, ourselves, are suddenly vulnerable to extinction. It doesn't get much more personal than that.

REVERSING EXTINCTION

It's not completely surprising that the idea of *de*-extinction—that we might be able to bring species that have gone extinct back to life—has attracted so much attention. If extinction is not forever, then it lets us off the hook. If we can bring species that we have driven to extinction back to life, then we can right our wrongs before it is too late. We can have a second chance, clean up our act, and restore a healthy and diverse future, before it is too late to save our own species.

While it is still not possible to bring extinct species back to life, science is making progress in this direction. In 2009, a team of Spanish and French scientists announced that a clone of an extinct Pyrenean ibex, also known as a bucardo, was born in 2003 to a mother who was a hybrid of a domestic goat and a different species of ibex. To clone the bucardo, the scientists used the same technology that had been used in 1996 to successfully clone Dolly the sheep. That technology requires living cells, so in April 1999, ten months before her death, scientists captured the last living bucardo and took a small amount of tissue from her ear. They used this tissue to create bucardo embryos. Only one of 208 embryos that were implanted into the surrogate mothers survived to be born. Unfortunately, the baby bucardo had major lung deformity and suffocated within minutes.

In 2013, Australian scientists announced that they successfully made embryos of an extinct frog—the Lazarus frog—by injecting nuclei from Lazarus frog cells that had been stored in a freezer for forty years into a donor cell from a different frog species. None of the Lazarus frog embryos survived for more than a few days, but genetic tests confirmed that these embryos did contain DNA from the extinct frog.

The Lazarus frog and bucardo projects are only two of the several de-extinction projects that are under way today. These two projects involve using frozen material that was collected prior to extinction and, consequently, are among the most promising of the existing de-extinction projects. Other de-extinction projects, including mammoth and passenger pigeon de-extinction, face more daunting challenges, of which finding well-preserved material is only one. These projects are proceeding nonetheless and, in the case of the mammoth, along several different trajectories. Akira Iritani of Japan's Kinki University is trying to clone a mammoth using frozen cells and claims that he will do so by 2016. George Church at Harvard University's Wyss Institute is working to bring the mammoth back by engineering mammoth genes into elephants. Sergey Zimov of the Russian Academy of Science's Northeast Science Station worries less about about how mammoths will be brought back than about what to do with them when it happens. He established Pleistocene Park near his home in Siberia and is preparing his park for the impending arrival of resurrected mammoths.

Not all de-extinction projects take a species-centric view. George Church's project is focusing on reviving mammoth-like traits in elephants, for example. While the goal of this project is to create an animal that is mammoth-like, its motivation is to reintroduce elephants into the Arctic. Stewart Brand and Ryan

Phelan have taken an even more holistic view. Together, they created a nonprofit organization called Revive & Restore, and are asking people to consider all the ways in which de-extinction and the technology behind it might change the world over the next few decades or centuries. In addition to initiating the passenger pigeon de-extinction project, Revive & Restore is driving several projects to revive living species that have dangerously low amounts of genetic diversity. With Oliver Ryder of San Diego's Frozen Zoo, for example, Revive & Restore is isolating DNA from archived remains of black-footed ferrets, which are nearly extinct in the present day. They hope to identify genetic diversity that was present in black-footed ferrets prior to their recent decline and, using de-extinction technologies, to engineer this lost diversity back into living populations.

In March 2013, Revive & Restore organized a TEDx event at National Geographic's headquarters in Washington, DC, to focus on the science and ethics of de-extinction. This media event was the first attempt to address de-extinction at a more sophisticated level than attention-grabbing headlines. When the event concluded, pubic opinion about de-extinction was mixed. Some people loved and others hated the idea that extinctions might be reversed. Fears were expressed about the uncertain environmental impacts of reintroduced resurrected species. Some ethicists argued that de-extinction is morally wrong; others insisted that it is morally wrong *not* to bring things back to life, if indeed it were possible to do so. Voices were also raised in opposition to the cost of de-extinction and whether the potential benefits justified this cost. What was lost in the noise of the ensuing public debates, however, was discussion of the current state of the science of de-extinction: what is possible now, and what will ever be possible? And, perhaps more importantly, there was little conversation and certainly no consensus about what the goal of de-extinction should be. Should we focus on bringing species back to life or on resurrecting extinct ecosystems? Or should the focus be on preserving or invigorating ecosystems in the present day? Also, and importantly, what constitutes a successful de-extinction?

In this book, I aim to separate the science of de-extinction from the science fiction of de-extinction. I will describe what we can and cannot do today and how we might bridge the gap between the two. I will argue that the present focus on bringing back particular species—whether that means mammoths, dodos, passenger pigeons, or anything else—is misguided. In my mind, de-extinction has a place in our scientific future, but not as an antidote to extinctions that have already occurred. Extinct species are gone forever. We will never bring something back that is 100 percent identical—physiologically, genetically, and behaviorally identical—to a species that is no longer alive. We can, however, resurrect some of their extinct traits. By engineering these extinct traits into living organisms, we can help living species adapt to a changing environment. We can reestablish interactions between species that were lost when one species went extinct. In doing so, we can revive and restore vulnerable ecosystems. This—the resurrection of ecological interactions—is, in my mind, the real value of de-extinction technology.

A SCIENTIFIC VIEW OF DE-EXTINCTION

I am a biologist. I teach classes and run a research laboratory at the University of California, Santa Cruz. My lab specializes in a field of biology called "ancient DNA." We and other scientists working in this field develop tools to isolate DNA sequences from bones, teeth, hair, seeds, and other tissues of organisms that used to be alive and use these DNA sequences to study ancient populations and communities. The DNA that we extract from these remains is largely in terrible condition, which is not surprising given that it can be as old as 700,000 years.

During my career in ancient DNA, I have extracted and studied DNA from an assortment of extinct animals including dodos, giant bears, steppe bison, North American camels, and saber-toothed cats. By extracting and piecing together the DNA sequences that make up these animals' genomes, we can learn nearly everything about the evolutionary history of each individual animal: how and when the species to which it belonged first evolved, how the population in which it lived fared as the climate changed during the ice ages, and how the physical appearance and behaviors that defined it were shaped by the environment in which it lived. I am fascinated and often amazed by what we can learn about the past simply by grinding up and extracting DNA from a piece of bone. However, regardless of how excited I feel about our latest results, the most common question I am asked about them is, "Does this mean that we can clone a mammoth?"

Always the mammoth.

The problem with this question is that it assumes that, because we can learn the DNA sequence of an extinct species, we can use that sequence to create an identical clone. Unfortunately, this is far from true. We will never create an identical clone of a mammoth. Cloning, as I will describe later, is a specific scientific technique that requires a preserved *living* cell, and this is something that, for mammoths, will never be found.

Fortunately, we don't have to clone a mammoth to resurrect mammoth traits or behaviors, and it is in these other technologies that de-extinction research is progressing most rapidly. We could, for example, learn the DNA sequence that codes for mammoth-like hairiness and then change the genome sequence of a living elephant to make a hairier elephant. Resurrecting a mammoth trait is, of course, not the same thing as resurrecting a mammoth. It is, however, a step in that direction.

Scientists know much more today than was known even a decade ago about how to sequence the genomes of extinct species, how to manipulate cells in laboratory settings, and how to engineer the genomes of living species. The combination of these three technologies paves the way for the most likely scenario of de-extinction, or at least the first phase of de-extinction: the creation of a healthy, living individual.

First, we find a well-preserved bone from which we can sequence the complete genome of an extinct species, such as a woolly mammoth. Then, we study that genome sequence, comparing it to the genomes of living evolutionary relatives. The mammoth's closest living relative is the Asian elephant, so that is where we will start. We identify differences between the elephant genome sequence and the mammoth genome sequence, and we design experiments to tweak the elephant genome, changing a few of the DNA bases at a time, until the genome looks a lot more mammoth-like than elephant-like. Then, we take a cell that contains one of these tweaked, mammoth-like genomes and allow that cell to develop into an embryo. Finally, we implant this embryo into a female elephant, and, about two years later, an elephant mom gives birth to a baby mammoth.

The technology to do all of this is available today. But what would the end product of this experiment be? Is making an elephant whose genome contains a few parts mammoth the same thing as making a mammoth? A mammoth is more than a simple string of As, Cs, Gs, and Ts—the letters that represent the nucleotide bases that make up a DNA sequence. Today, we don't fully understand the complexities of the transition from simply stringing those letters together in the correct order—the DNA sequence, or genotype—to making an organism that looks and acts like the living thing. Generating something that looks and acts like an extinct species will be a critical step toward successful de-extinction. It will, however, involve much more than merely finding a well-preserved bone and using that bone to sequence a genome.

When I imagine a successful de-extinction, I don't imagine an Asian elephant giving birth in captivity to a slightly hairier elephant under the close scrutiny of veterinarians and excited (and quite possibly mad) scientists. I don't imagine the spectacle of this exotic creature in a zoo enclosure, on display for the gawking eyes of children who'd doubtless prefer to see a *T. rex* or *Archaeopteryx* anyway. What I do imagine is the perfect arctic scene, where mammoth (or mammoth-like) families graze the steppe tundra, sharing the frozen landscape with herds of bison, horses, and reindeer—a landscape in which mammoths are free to roam, rut, and reproduce without the need of human intervention and without fear of re-extinction. This—building on the successful creation of one individual to produce and eventually release entire populations into the wild—constitutes the second phase of de-extinction. In my mind, de-extinction cannot be successful without this second phase.

The idyllic arctic scene described above might be in our future. However, before a successful de-extinction can occur, science has some catching up with the movies to do. We have yet to learn the full genome sequence of a mammoth, for example, and we are far from understanding precisely which bits of the mammoth genome sequence are important to make a mammoth look and act like a mammoth. This makes it hard to know where to begin and nearly impossible to guess how much work might be in store for us.

Another yet-to-be-solved problem is that some important differences between species or individuals, such as when or for how long a particular gene is turned on during development or how much of a particular protein is made in the gut versus in the brain, are inherited epigenetically. That means that the instructions for these differences are not coded into the DNA sequence itself but are determined by the environment in which the animal lives. What if that environment is a captive breeding facility? Baby mammoths, like baby elephants, ate their mother's feces to establish a microbial community capable of breaking down the food they consumed. Will it be necessary to reconstruct mammoth gut microbes? A baby mammoth will also need a place to live, a social group to teach it how to live, and, eventually, a large, open space where it can roam freely but also be safe from poaching and other dangers. This will likely require a new form of international cooperation and coordination. Many of these steps encroach on legal and ethical arenas that have yet to be fully and adequately defined, much less explored.

Despite this somewhat pessimistic outlook, my goal for this book is not to argue that de-extinction will not and should never happen. In fact, I'm nearly certain that someone will claim to have achieved de-extinction within the next several years. I will argue, however, for a high standard by which to accept this claim. Should de-extinction be declared a success if a single mammoth gene is inserted into a developing elephant embryo and that developing elephant survives to become an adult elephant? De-extinction purists may say no, but I would want to know how inserting that mammoth DNA changed the elephant. Should de-extinction be declared a success if a somewhat hirsute elephant is born with a cold-temperature tolerance that exceeds that of every living elephant? What if that elephant not only looks more like a mammoth but is also capable of reproducing and sustaining a population where mammoths once lived? While others will undoubtedly have different thresholds for declaring de-extinction a success than I do, I argue that this—the birth of an animal that is capable, thanks to resurrected mammoth DNA, of living where a mammoth once lived and acting, within that environment, like a mammoth would have acted—is a successful de-extinction, even if the genome of this animal is decidedly more elephant-like than mammoth-like.

MAKING DE-EXTINCTION HAPPEN

Many technical hurdles stand in the way of de-extinction. While science will eventually find a way over these hurdles, doing so will require significant investment in both time and capital. De-extinction will be expensive. There will be important issues to consider about animal welfare and environmental ethics. As with any other research project, the cost to society of the research needs to be weighed against the gains to society of what might be learned or achieved.

If we brought back a mammoth and stuck it in a zoo, then we could study how mammoths are different from living elephants and possibly learn something about how animals evolve to become adapted to cold climates. Some scientists who favor de-extinction see this as a reasonable goal, and many nonscientists would be just as happy to see unextinct species in zoos as they would be to see them in safari parks or unmanaged wild habitat. But is bringing a mammoth back to life so that we can look at it and possibly study it enough of a societal gain to justify the costs of creating that mammoth?

If, like elephants, mammoths helped to maintain their own habitat, then bringing mammoths back to life and releasing them into the Arctic may transform the existing tundra into something similar to the steppe tundra of the ice ages. This might create habitat for living and endangered Arctic species, such as wild horses and saiga antelopes, and other extinct megafauna that might be targets for de-extinction, such as short-faced bears. Is the possibility of revitalizing modern habitats in a way that benefits living species enough to justify the expense? Of course, ecosystems change and adapt over time, and there is no certainty that the modern tundra would convert back to the steppe tundra of the Pleistocene even with free-living populations of unextinct mammoths. Should uncertainty of success influence our analysis of the cost of de-extinction?

What if we identified a very recently extinct species that played a similarly important role in a present-day environment, and brought that species back to life? For example, kangaroo rats are native to the deserts of the American Southwest, but their populations have become increasingly fragmented over the last fifty years, and many subspecies are known to be extinct today. Kangaroo rats are so important to their ecosystem that their disappearance can cause a desert plain to turn into arid grassland in less than a decade. The domino effects of kangaroo rat extinction include the disappearance of plants with small seeds and their replacement by plants with larger seeds (on which the kangaroo rat would have fed), in turn leading to a decline in seed-eating birds. The decrease in foraging and burrowing slows plant decomposition and snowmelt, and the lack of burrows leaves many smaller animal and insect species without shelter. When the kangaroo rat goes extinct, the entire ecosystem is in danger of the same. If bringing the kangaroo rat back could save the entire ecosystem, would that be sufficient to justify the expense?

In the chapters that follow, I will walk through the steps of de-extinction. As I indicated earlier, de-extinction is likely to happen in two phases. The first phase includes everything up to the birth of a living organism, and the second will involve the production, rearing, release, and, ultimately, management of populations in the wild. For each step in the process, I will describe what we now know, what we need to know, what we are likely to know soon, and what's likely to remain unknown. I will discuss both the science and the ethical and legal considerations that are likely to be part of any de-extinction project. Although the book is organized as a how-to manual, de-extinction is not a strictly linear process, and not all steps will apply to every species. Species from which living tissue was cryopreserved prior to their extinction may be clonable in the traditional sense, for example, while other species will require additional steps to create a viable embryo.

As part of my professional relationship with Revive & Restore, I have been involved in research that focuses on two species—mammoths and passenger pigeons—that are presently targets of de-extinction efforts. This will no doubt result in an animal-centric (really mammoth and pigeon-centric) view of the process. Still, many of the details will be broadly applicable across taxonomic lines. My hope is to present a realistic but not cynical view of the prospects for de-extinction, which I believe has the potential to be a powerful new tool in biodiversity conservation.

CHAPTER 2

SELECT A SPECIES

I taught a class on the topic of de-extinction recently for graduate students studying ecology at UC Santa Cruz. For their first assignment, I asked each of the students to choose an extinct species that they'd like to see brought back to life and to become that species' de-extinction advocate. I was curious to learn not only which species they would choose but also what their motivations for choosing a species would be. Because the students were ecologists, I expected them to focus on the impact of each potential "de-extinctee" on the environment into which it would be released, should de-extinction be successful. They did not.

The students selected, among other species, the Yangtze River dolphin, the dodo, the moa, the Tasmanian tiger, the Cascade Mountains wolf, Steller's sea cow, and *Thismia americana*—a tiny, translucent plant that is so poorly described that it lacks a common name. Some of their arguments in favor of de-extinction were purely research-oriented—imagine what could be learned by studying this species—while others were more practical—imagine how this species might create new opportunities for ecotourism. Most students discussed the technical challenges of de-extinction—it would be hard to find well-preserved dodo remains, for example, or a surrogate mother for a Tasmanian tiger. Some students acknowledged that suitable habitat might be hard to find, and that existing laws might make it hard to protect the species once it was released into the wild. Few, however, discussed what effect introducing an unextinct species into an existing community might have, which surprised me.

As the class progressed, it became clear that the students had different motivations for selecting candidate species for de-extinction. Some students wanted to bring a species back simply because it would be exciting to do so. Others chose species that they believed could provide significant ecological and environmental benefits, or that might improve our understanding of the evolutionary processes that lead to diverse forms of life. One student selected the species that he believed had the fewest technical barriers in the way of success.

None of these are "wrong" reasons to choose a species for de-extinction. However, the diversity of motives within this small group highlights the first significant challenge faced by scientists doing on-the-ground de-extinction work: to agree on what to bring back. How do we decide which species should be the first targets of de-extinction? Should we choose the species that will be the easiest to bring back? The most awe-inspiring? The most likely to draw attention, perhaps motivating further investment into the technology? Or should we focus on those species whose de-extinction is clearly scientifically justifiable? And, if the latter, what does that mean exactly? Finally, and just as importantly, who is the "we" that gets to decide?

THE "RIGHT" REASONS FOR DE-EXTINCTION

As I suggest above, there are likely to be many reasons to select (or not to select) a particular species for de-extinction. Whether de-extinction is technically feasible, and whether suitable habitat exists into which a species might be reintroduced are important considerations. These questions address whether a species *can* be brought back, however, rather than whether it *should* be brought back. The latter is, unsurprisingly, a much more difficult question to answer.

Let's consider the Yangtze River dolphin, for example. Bringing back the Yangtze River dolphin would certainly be exciting, which some may feel is sufficient motivation to try. Those who would benefit most from its de-extinction would probably be more likely to advocate for it over other candidate species. But who would those people be? The students offered three substantive arguments in favor of bringing back the Yangtze River dolphin, each highlighting different potential benefits, and therefore beneficiaries, of Yangtze River dolphin de-extinction.

The demise of the Yangtze River dolphin—also known as the baiji—is a terribly sad case. My friend Sam Turvey, who works for the Zoological Society of London, has devoted a big part of his life to looking out for species that are on the brink of extinction. In 2006, he led an expedition to survey the Yangtze River for any signs of river dolphins. Sam and his team searched the Yangtze River system for two months and saw no dolphins or signs of dolphin life. With heavy hearts, they declared the Yangtze River dolphin functionally extinct.

The first of the students' arguments in support of Yangtze River dolphin de-extinction emphasized the evolutionary distinctiveness of the Yangtze River dolphin. Only two other freshwater dolphin species—the Ganges River dolphin of Southeast Asia and the Amazon River dolphin of South America—are known. When scientists first described the river dolphins, they noticed that the three species looked very much alike. All three river dolphins have long, narrow mouths, for example, with lots of teeth. They also have small eyes, compared with their marine relatives. Scientists decided that these morphological similarities probably meant that the three river dolphin species were descended from a single common ancestor species that was also a river dolphin. When genetic data became available, however, it was clear that this was not true. Instead of confirming a single evolutionary lineage, the genetic data indicated that each species made a separate transition from the ocean to freshwater. The morphological similarities among them arose from a combination of shared, deep ancestry and convergent evolution—life in similar environments led to the emergence of similar traits. This makes each freshwater dolphin species particularly valuable from a scientific perspective. We can compare their genomes, physiologies, and behaviors to better understand how species adapt to freshwater environments. Scientists would be one group to benefit from Yangtze River dolphin de-extinction.

The second of the students' arguments pointed out that rare things intrigue everyone, and not only scientists. If the Yangtze River dolphin is brought back to life, the spectacle of its existence would likely be sufficient to attract certain types of tourists who would be keen to see the animal firsthand. Ecotourism is one of the most rapidly growing sectors of the tourism industry. It both provides jobs and inspires communities to take advantage of local natural resources. Tourists would come to take photos, sleep in a local hotel, eat at a few local restaurants, and maybe even buy a stuffed toy dolphin replica to take home. Yangtze River dolphin de-extinction would have a positive economic impact. The people who live in the region of reintroduction would benefit, as would tourists—some may even be inspired to care just a little bit more about the plight of the native species back home.

The students' final argument postulated that Yangtze River dolphins should be brought back because their de-extinction would necessarily have a positive impact on the environment. The Yangtze River is presently too polluted to support dolphins, so this situation would have to change. Bringing back the river dolphin would require making the river ecosystem a cleaner, healthier one, with far-reaching ecological benefits.

This same multi-faceted rationale holds true for other species. For example, another group of animals that my students felt were good candidates for de-extinction were the moa of New Zealand. As with the Yangtze River dolphin, the reasons for bringing back the moa are both scientific—moa have no close living relatives, so understanding their biology and behavior is nearly impossible without bringing them back—and economic—living moa would provide yet another reason for people to visit New Zealand, which is already a popular ecotourism destination. Resurrected moa may also re-establish missing interactions with other species, benefiting the native ecosytems of New Zealand.

Moa were enormous birds that did not fly (figure 1). Some

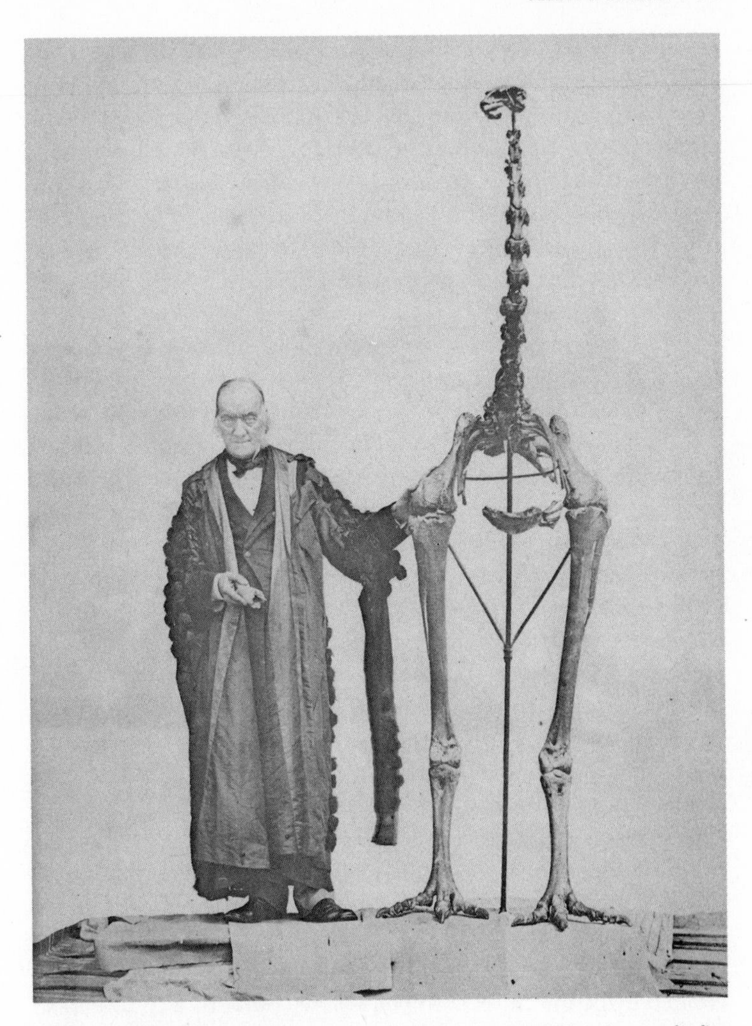

Figure 1. Sir Richard Owen and his reconstruction of a giant moa, *Dinornis novazealandiae*. In his right hand, Owen holds the first moa bone that he examined. This photograph was first published in Owen's book, *Memoirs on the Extinct Wingless Birds of New Zealand*, vol. 2 (London: John van Voorst, 1879). Courtesy of the University of Texas Libraries, The University of Texas at Austin.

species of moa reached more than three meters tall with their necks outstretched and weighed more than two hundred kilograms. Because they didn't fly, moa were easy targets for the first inhabitants of New Zealand—the Māori—who hunted them for food, used their bones to make jewelry and fishing gear, and fashioned their skins and feathers into clothing. Māori and moa coexisted on the islands of New Zealand for more than three hundred years before hunting and habitat loss eventually led to the moa's extinction.

In New Zealand, moa are a symbol of national pride. Very briefly in the 1890s, New Zealand was officially dubbed the "Land of the Moa," thanks in part to a play of the same name written by George Leitch. New Zealanders have created moa artwork, moa poetry, and even moa beer, and many New Zealanders are strongly in favor of bringing moa back to life. New Zealand has a strong record of environmentalism and protection of native species and habitat, which means that unextinct moa would probably be provided with a safe place to live should they be resurrected. However, some of the challenges of de-extinction mean that unextinct moa would probably not be 100 percent identical to the species that once inhabited New Zealand but rather genetic hybrids with nonnative birds. It is not clear how these hybrid animals might be accommodated within the environmental philosophies of many New Zealanders.

A third popular choice among my students was the dodo (figure 2). Dodos were large, flightless pigeons that were endemic to Mauritius, a volcanic island situated in the Indian Ocean about 1,200 miles from the southeastern coast of Africa. In 1507, Portuguese sailors landed on Mauritius, which was at that time uninhabited by people, after being blown off course by a cyclone. The Portuguese were not particularly interested in the island and did not establish a permanent colony. Dutch sailors arrived about ninety years later but didn't stick around either. They did, however record for the first time a large, fat, flightless bird with little to no fear of humans. In 1638, twenty-five Dutch sailors returned to Mauritius and established the first permanent human settlement. Twenty-four years later, the dodo was extinct. Based on

Figure 2. Dodo, *Raphus cucullatus*. Illustration by Adrian van den Venne, probably around 1626.

written accounts of the interactions between humans and dodos, humans are clearly to blame for the dodo's extinction.

As with the Yangtze River dolphin and the moa, scientific, ecological, and economic interests can be cited as reasons to resurrect the dodo. The dodo is a large, flightless pigeon whose closest relative is a small, strong flyer. Studying its genome could help scientists to better understand how traits such as flightlessness and gigantism evolve. Reintroducing dodo populations to Mauritius would require the creation of suitable habitat, which would mean removing invasive species and establishing new protected zones, which would benefit both the local people and the native ecosystem. The dodo is a special case for de-extinction, however, because it, more than any other species, is the international symbol of human-caused extinctions. If candidate species were to be ranked according to the potential psychological impact of their de-extinction, the dodo would be very high on the list.

A SHORT GUIDE TO DE-EXTINCTION DECISION MAKING

The three examples above highlight what I have come to understand as a general principle of selecting de-extinctees. Most people are at least somewhat uncomfortable with the idea of de-extinction. However, when forced to come up with *one* suitable candidate species, nearly everyone chooses something that is extinct because humans made it so. The Yangtze River dolphin is extinct because we destroyed its habitat. Moa are extinct because we hunted them to death. Dodos are extinct because we introduced cats and rats and pigs to Mauritius, and these cats and rats and pigs made easy meals of all the dodo eggs they could find. If it had not been for humans, each of these species would probably still be alive.

In addition to how a species became extinct, other characteristics of species make them more or less popular choices for de-extinction. Perhaps unsurprisingly, most people would prefer to resurrect herbivores, rather than carnivores. Less obviously, most people choose large species for de-extinction rather than small species, presumably because large species are, well, bigger. And most people choose to bring back animals, as opposed to plants, fungi, or any other living thing.

It is enormously important to make an informed rather than emotional decision about whether an extinct species should be brought back to life. Different species require different technical innovations, different amounts of hands-on manipulation, and different habitat. Some species would be decidedly easier to bring back than others. Some species would proceed relatively straightforwardly through the early steps of de-extinction, such as sequencing their genome, but present potentially insurmountable challenges during later steps, such as identifying suitable habitat into which they could be released. When considering whether a species is a good candidate for de-extinction, it is tempting to focus only on those steps leading to the birth of a newly unextinct animal and to ignore the later steps of rearing and reintroduction into the wild. It is unwise, however, and I would go so far as to say unfair, to proceed through these first steps without careful evaluation of the entire process from fertilized egg to free-living population. What is the point, after all, of bringing a species back from the dead if it is not to reestablish a wild population?

To simplify navigation through the process of selecting a species for de-extinction, I propose seven questions that should be asked and answered. The questions fall into two broad categories. The first set of questions attempts to place that species in the context of its ecosystem. How did it interact with and affect other species when it was alive, and how might that be different today? The second set of questions turns to the nitty-gritty of the science. Is de-extinction of this species—or at least of some specific traits that defined this species—practical, given current and future technologies? My focus for now is on the technical aspects of de-extinction and not the ethical questions that will undoubtedly arise throughout the process. Obviously, these questions are not exhaustive, and not every question applies to every species. These questions do, however, provide a useful means to think through the implications of de-extinction and, perhaps, avoid some potential disasters.

Is There a Compelling Reason to Bring This Species Back?

Probably—and hopefully—the first question that comes to the minds of most people when contemplating de-extinction is "Why?" Why *this* species? Why *now*? Why *here?* As I noted before, most people advocate for species that they know, beyond reasonable doubt, are extinct because of something that humans have done. Bringing these species back goes some way to mitigate the guilty conscience of an ecological savvy human. But mitigation of guilt is not a compelling reason to bring something back to life. I may feel some guilt by association with my Native American ancestors, who probably participated in hunting short-faced bears, and with my European ancestors, who probably were in some way involved with the extinction of the Neandertals. This does not mean I want to bring back short-faced bears and Neandertals. Indeed, in both of these cases, de-extinction for the purposes of alleviating guilt seems remarkably selfish; what kind of existence would either of these have in the world today?[1]

Compelling reasons to bring something back to life are more likely to relate to the species themselves and the roles these species are likely to play in the environment of the present day. For example, if the species filled a particularly important niche within its ecosystem, then its loss is likely to have resulted in chaotic destabilization of that ecosystem. Bringing it back might restore lost interactions between species and restabilize the ecosystem, in turn saving other species from extinction. The kangaroo rats that I mentioned earlier are a good example of keystone species that play important, stabilizing roles in their ecosystems. The Cascade Mountains wolf—an additional species suggested for de-extinction by a student in my class—is another. Importantly, both of these are very recent extinctions, and their ecosystems may not yet have adapted to accommodate their loss.

The potential role of the Cascade Mountains wolf in maintaining ecological balance can be extrapolated from work in Yellowstone National Park over the last two decades. When wolves were reintroduced to Yellowstone National Park in 1995, many people were convinced this was going to lead to disaster. The cause for concern was that wolves are predators and, as such, would likely depredate livestock from local farms, much to the dismay of the ranchers who depend on the livestock. This was an appropriate concern. As wolf populations have grown, there have been many instances of wolves taking livestock. However, the main source of the wolves' diet is local wildlife, and elk in particular. By 2006, the elk population in Yellowstone had shrunk to 50 percent of the size it had been when wolves were first reintroduced to the park. Today, elk are no longer overgrazing the plants and young trees that grow along the meadows and valley bottoms, and consequently, woody plants are making a comeback throughout the park. The increase in woody plants provides a greater diversity of habitat for small mammals, whose populations are also on the rebound. Wolves are outcompeting coyotes, whose populations had become much larger after the disappearance of wolves. Fewer coyotes is good news for the animals that coyotes like to eat, including red foxes, pronghorn, and sheep.

Certainly, wolves are predators. Wolves will take livestock when they have the opportunity to do so. However, it seems clear that restoring wolves to Yellowstone National Park has played a critical role in stabilizing the Yellowstone ecosystem.

The Cascade Mountains wolf is a subspecies of gray wolf that lived in the mountains of Washington, Oregon, and British Columbia until around 1940. Based on the positive results of the Yellowstone wolf reintroduction, there is compelling ecological reason to bring back the Cascade Mountains wolf and reintroduce it to its former range.

The case of the Cascade Mountains wolf touches on another intriguing issue. This wolf was a subspecies of gray wolf, and not its own, distinct species. This raises a different question: is it appropriate to select a *subspecies* for de-extinction?

Before attempting to answer that question, I should first clarify what it means to be a subspecies as opposed to a species or population. From an ecological perspective, a *population* is a group of individuals of the same species that live together in the same place. The individuals interbreed, compete with each other for resources, and share the same geographic space. A *species* tends to be defined as an evolutionary lineage that is reproductively isolated from all other evolutionary lineages. Individuals of the same species can move between populations with little consequence to their ability to find a mate and reproduce. Individuals belonging to different species cannot mate. Or if they do, the offspring that are born either do not survive into adulthood or cannot have offspring themselves.

This species-as-reproductively-isolated-lineages concept, known as the biological species concept, was formally described by Ernst Mayr in 1942. The concept turns out to have some flaws. Specifically, some lineages that we strongly believe are separate species are not strictly reproductively isolated. Polar bears and brown bears, for example, are commonly considered to be two different species. But bears born from crosses between brown bears and polar bears survive and can continue to mate and produce offspring. Dogs, wolves, and coyotes can and do interbreed frequently. Cows and bison and yaks can all interbreed and produce fertile offspring. And ancient DNA from Neandertal bones revealed that our species can (and did) mate with Neandertals and that, as a result of this hybridization, Neandertal genes survive in all living humans with Asian or European ancestry.

Why do biologists hold on to this confusing system? As humans, we are compelled to categorize. When we see chaos, we desire to transform that chaos into something ordered so that our brains can make sense of it. Clearly, evolution does not work in absolutes. An animal is not born one day as an entirely new species, incapable of reproducing with anyone in its parents' species. Instead, speciation is a long process involving many underlying genetic and behavioral changes. Populations become geographically isolated and evolve along independent trajectories. Eventually, enough changes will have evolved so that individuals are incapable of breeding between populations. As we see with brown bears and polar bears and with humans and Neandertals, however, what common sense would call species-level differences will sometimes evolve before the two lineages are completely reproductively isolated.

To impose order on the disorder that is biology, Carl Linnaeus, an eighteenth-century Swedish biologist and physician, devised a taxonomic system to describe and categorize all forms of life. His system provides a hierarchical classification of everything according to its relationship with everything else. The biggest bins classify organisms into kingdoms: Animalia, Plantae, Fungi, Protista, Eubacteria, and Archaeobacteria (although the latter two are sometimes grouped into one kingdom, the Monera). Wolves, dogs, bears, snakes, and rabbits are all animals, so they all belong to the kingdom Animalia. Within that, wolves, coyotes, bears, and rabbits are mammals (class Mammalia). Wolves, coyotes, and bears are carnivores (order Carnivora). Wolves and coyotes are canids (family Canidae). Both belong to the genus *Canis*, but wolves are *Canis lupus* and coyotes are *Canis latrans*, where *lupus* and *latrans* are the official Latin names for the two different species.

After that, it gets messy. Sometimes species are subdivided into *subspecies*. But this is tricky. Some taxonomists will refer to a population that seems to be particularly isolated from other populations as a subspecies, while a different taxonomist might look at the same population and decide that it is not sufficiently different to merit subspecific status. Unlike a species, there's really no rule to go by to decide whether a subspecies is real or not.

What does all of this have to do with de-extinction? A lot. If a subspecies is not real, or if it is just a slightly different version of something that is not extinct, should time and energy be spent to bring that subspecies back to life?

Sometimes subspecies are defined geographically. This means that, while there may be no physical or genetic barrier to interbreeding, they are simply too far apart in geographic terms for interbreeding to take place. For example, it is not particularly likely that the Iberian wolf will breed with the Mexican wolf. In the absence of interbreeding, the two populations will accumulate genetic differences that make them look and act differently from each other. There is no doubt, however, that both of these are *wolves*. So, if either Mexican wolves or Iberian wolves were to go extinct, would it be reasonable to use de-extinction technology to bring them back?

Consider a hypothetical scenario in which there are two subspecies that are ecologically very important in their ecosystems—keystone species—and one of these goes extinct, destabilizing the ecosystem in which it lived. The two subspecies are very closely related to each other. In fact, the only things that differentiate them are that they lived in different places and had some small morphological difference—let's say they had slightly differently shaped ears. To restabilize the ecosystem, we plan to reintroduce the extinct keystone species. Is it better to bring it back to life using de-extinction science or to introduce the close relative? In other words, how different does an extinct lineage have to be from a living lineage to justify its de-extinction?

From a technical perspective, de-extinction of a subspecies like the Cascade Mountains wolf would be much simpler than de-extinction of a distinct species. As I will discuss, assembling the genome sequence of an extinct organism can be extremely challenging and requires a guide genome to act as a scaffold onto which the short, damaged fragments of ancient DNA can be mapped. The Cascade Mountains wolf genome could be assembled using another gray wolf genome as a guide, simplifying this process greatly. Cascade Mountains wolf embryos could be implanted into a mother gray wolf, and families of gray wolves could rear the pups in established gray wolf packs. This then raises the question, how would the purportedly unextinct Cascade Mountains wolf differ from the wolf subspecies into which it is born? Would it be preferable simply to introduce another gray wolf subspecies into the Cascade Mountains?

While some species or subspecies seem too similar to living species to justify their de-extinction, other extinct species have no evolutionarily close living relatives. This argues both *for* their de-extinction, because bringing them back will restore more evolutionary novelty than would bringing back something that has a close, living relative, and *against* their de-extinction, because bringing them back will be much more costly to achieve.

Moa, for example, were subdivided into three extinct families within the order Dinornithiformes, which has no living representative. The closest living relative to the moa is the tinamou, and the common ancestor of moa and tinamou lived around 50 million years ago. The moa represents a long history of independent evolution, and bringing it back would restore many unique traits to the world. However, with no close relative, it would be extremely hard to assemble the broken bits of DNA recovered from moa bones into a reasonably accurate moa genome. In this case, the guide genome would be more than 100 million years diverged—twice the evolutionary distance to the common ancestor—from the ancient genome. The same is true for any extinct species that lacks an evolutionarily close living relative. Identifying appropriate surrogate mothers or eggs within which the embryos could develop would be extremely challenging for species that lack close living relatives. We would also have little way to know what the native behaviors of such species should be, how much parental care would be required to rear them, or how to mimic this parental care or other important social interactions. In a sense, these individuals might be too different from anything that is alive for de-extinction to succeed.

The ideal candidate for de-extinction has both sufficiently closely related living relatives to make its de-extinction feasible and unique traits or adaptations to a particular habitat. The golden toad, for example, was last seen in the cloud forests of Costa Rica in 1989. It was such a peculiar, bright orange color that Jay Savage, the herpetologist who described it, had trouble believing it was real and not some trick. The golden toad was tiny—adult males measured a bit more than five centimeters in length—and is a good candidate for de-extinction. It belongs to the genus *Bufo,* which is species-rich and diverse and therefore has many close relatives that are not extinct. However, among its many close relatives, the golden toad was the only *Bufo* to display such a striking orange color. What if the proteins that made the orange color had some undiscovered medical purpose, or psychoactive properties? We'll never know until somebody licks one, and for that we'll have to bring it back.

Finally, the ideal candidate for de-extinction may be one that has only recently gone extinct. Ecosystems are constantly in flux. They are influenced by abiotic changes, such as how much rain falls or how cold the winters are, and biotic changes, including both species extinctions and introductions. When a species goes extinct, the ecosystem in which it lived changes to adapt to its disappearance. If the extinction took place many thousands or possibly even only several hundred years ago, it may be that reintroducing that species would actually destabilize whatever new equilibrium that ecosystem had achieved. This does not mean that the only acceptable de-extinctions will be of recently extinct species. Certainly, some species, such as large herbivores, played roles in ancient ecosystems that have not been filled in their absence. As I discuss below, it may be that careful evaluation of the relative risks and rewards of their de-extinction will lead to the conclusion that they, too, should be brought back on the grounds of their potential beneficial impact on the environment of the present day.

Why Did This Species Go Extinct the First Time?

People tend to be most interested in bringing back species that went extinct because of human activities. Given that proposition, asking the question "Why did they go extinct?" seems a bit silly. In fact, it is not at all silly. Humans are remarkably creative when it comes to killing things.

We killed many species using brute force. In the nineteenth century, we netted and shot billions of passenger pigeons, eventually leading to their extinction (figure 3). When Europeans arrived in North America, passenger pigeons were thought to make up 25–40 percent of the bird population in the eastern United States. In 1866, a report described a single flock of more than 3.5 billion passenger pigeons flying across the Ohio skies, taking

Figure 3. A flock of migrating passenger pigeons. Drawing from *The Illustrated Shooting and Dramatic News*, July 3, 1875. ©The Archives and Manuscripts Department, John B. Cade Library, Southern University and A&M College.

more than fourteen hours to pass overhead. At one o'clock in the afternoon on the first of September 1914, the last remaining passenger pigeon, named Martha, died in captivity in the Cincinnati Zoo (plate 1).

Extinction due to overexploitation is a common theme in human history, and a human tendency that we are still grappling to overcome. Steller's sea cow was a large—nine meters long and up to ten tons in weight—marine mammal whose closest living relative is the dugong (figure 4). Steller's sea cows were once abundant throughout the North Pacific but were hunted to death after their discovery in the eighteenth century. Overexploitation is also blamed for the extinction of the great auk, which was taken for its fat, feathers, meat, and oil. We continue to overexploit the species we rely on today. In 2012, the *State of the World Fisheries and Aquaculture Report* reported that 30 percent of the

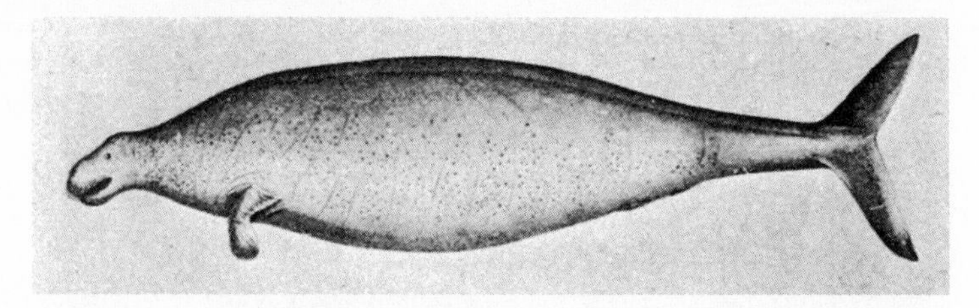

Figure 4. Steller's sea cow, *Hydrodamalis gigas*. Illustration by J. F. Brandt, 1846. This image was published in *Extinct Animals*, by E. R. Lankester (London: A. Constable, 1905).

world's fisheries were overexploited and would require strict management to be sustainable into the future.

But we don't only kill things with brute force. Indirect effects of human population growth—including the conversion of wild habitat for cities, towns, and agricultural land; deforestation; monoculture; and the construction of roads and highways to connect all of these things—result in changes to habitats that disrupt and destabilize ecosystems, leading to extinctions. Bird species are particularly susceptible to habitat destruction; the gradual clearing of forests in order to make way for people and agriculture on the islands of the Pacific alone has been blamed for the extinction of thousands of bird species. Indeed, habitat destruction accounts for more than half of the endangered birds of the world today.

We also move around the world, and as we move we bring stuff with us, both accidentally and intentionally. We introduce parasites, predators, and competitors to ecosystems in which these organisms did not exist previously, resulting in extinctions. Again, birds on islands are particularly susceptible to introduced predators, especially rats, cats, and snakes, which are keenly adept at finding and consuming eggs. Rats and cats are blamed for the extinction of the Tahitian sandpiper and Society parakeet in the Society Islands; the dodo, solitaire, Reunion pigeon, Rodrigues starling, and red rail in the Mascarenes; and the Raiatea parakeet, white-winged sandpiper, and Maupiti monarch in French Polynesia, to name only a few. In addition to introduced predators, diseases brought to these new locations spread from domestic to wild species, also causing extinctions.

And, with the byproducts of agriculture and industry, we pollute the world around us. The Yangtze River dolphin is an example of how pollution, combined with habitat destruction, can lead to extinction. The Madeiran large white butterfly was declared extinct recently, its extinction blamed on a combination of habitat destruction and pollution from agricultural fertilizers.

In some cases, even when it is clear that humans are ultimately at fault for an extinction, it is very difficult to know what the proximate causes of that extinction were. If the cause of the extinction is not completely understood, the species is most likely not a good candidate for de-extinction. If we don't know what caused a species to go extinct the first time, how can we know that it won't quickly become extinct again? Equally, some species for which we do know the most proximate cause of extinction are poor candidates for de-extinction. For example, despite the emotional allure of dodo de-extinction, if we were to bring dodos back and reintroduce them to Mauritius, their eggs would be promptly gobbled up by the large rat and cat populations that thrive on the island today.

Is There a Place for This Species to Live if We Successfully Bring It Back?

If we know why the species we are considering for de-extinction went extinct in the first place, is it possible to correct whatever went wrong? In the case of the dodo, we would have to create a cat- and rat-free zone on Mauritius into which the unextinct dodos could be placed. If this is not possible, either because there is no habitat that could be set aside for these purposes or because it's just too hard to keep out rats and cats, then the dodo is not an appropriate candidate for de-extinction.

The loss of suitable habitat—whether by deforestation, development, pollution, or the introduction of parasites and predators—seems to be the most common cause of extinctions that have been caused indirectly by humans (if we assume that humans only cause extinctions directly by overexploitation). Yet, more and more humans live on this planet every day, which means we take up more and more space and need more and more food, which translates into the increasingly rapid and destructive conversion of natural habitat into land for human use. A major challenge to any de-extinction effort will therefore be to secure suitable habitat for the unextinct species. Suitable habitat will have to (i) include appropriate prey or food to sustain the unextinct population; (ii) exclude predators or competitors (including invasive species) that would drive the species back into extinction, while at the same time leaving sufficient numbers of carnivores in the system so as not to destabilize the food web; (iii) be devoid of disease, parasites, or pollutants; (iv) imitate as much as possible the temperature and precipitation regimes of the native habitat of the species; and (v) be sufficiently large to support a self-sustaining population.

Interestingly, this whole process might be simpler for species that we drove to extinction directly through hunting, as their habitat may still exist. Of course, the survival of these species would rely on our not hunting them to death a second time. Like exotic species that are alive today, unextinct species would have to be protected from increasingly creative and dangerous poachers, using laws and statutes that are in many cases difficult to enforce.

How Will Introducing This Species Affect the Existing Ecosystem?

The extinction of a species changes the ecosystem in which it once lived. Over time, the ecosystem restabilizes, and the niche that was once filled by the extinct species is either filled by a different species or eliminated. The longer a species has been gone, the more likely it is that the ecosystem has adapted to its absence. So when the species is reintroduced, what effect will that reintroduction have on the species that live there today?

When passenger pigeons flocked in large numbers across eastern North America, the landscape was different than it is today. The deciduous forest was more widespread, American chestnut trees were plentiful, and people were not. Passenger pigeons were a dominant and destructive force in the eastern deciduous forest. They fed primarily on large seeds: oak acorns and the nuts of hickories, beeches, and chestnuts. Flocking in the billions, hungry passenger pigeons could destroy the entire seed crop of a forest stand in very little time. When they nested, as many as five hundred birds would nest in a single tree, and when they left the nests, they tended to leave behind dead trees covered in bird droppings. When they went extinct, this avian version of a nonstop EF5 tornado came to a screeching (squawking?) halt. Since that time, humans have converted much of the historic deciduous forest into towns, cities, and agricultural land. What would a flock of a billion passenger pigeons eat today? What effect would their de-extinction have on the remaining deciduous forest? On agriculture? On the bird species and other animal species that live there today, with which unextinct passenger pigeons would compete for access to food and nesting grounds?

It may be that some unextinct species would have minimal destabilizing effects on present-day ecosystems. Careful evaluation of the consequences of reintroduction is required, however, and not simply from the perspective of whether that particular species would survive after reintroduction. If a species' de-extinction would lead to changes to the existing habitat that in turn would threaten living species, then that species is not a good candidate for de-extinction.

Finally, as disingenuous as it may seem at first, it would be remiss not to think about the effect that the de-extinction would have on human populations. Few East Coasters would be overjoyed at the site of a billion passenger pigeons darkening the sky just above their freshly manicured lawns and newly waxed SUVs, for example. But there are subtler aggravations that would probably make this particular de-extinction unpopular. If unextinct passenger pigeons were protected as endangered species, people who enjoy hunting pigeons for sport may find themselves facing new restrictions about when and where pigeons can be hunted, or even whether pigeons can be hunted at all, given the presumed difficulty of distinguishing the unextinct passenger pigeons from common pigeons that were once unprotected game. And a billion passenger pigeons would probably require quite a lot of protected habitat, which would have to be repurposed from somewhere.

These problems of course extend beyond passenger pigeons. If new and different categories of protected species emerge, regulations enacted to prevent their (re)extinction may make previously accessible wilderness off-limits to recreation, much to the chagrin of hunters, campers, hikers, and so on. Farmers are unlikely to support de-extinction of species like the Carolina parakeet, which was driven to extinction precisely because it was an agricultural pest. And ranchers already unhappy about the idea of wolf reintroduction will hardly embrace the idea of saber-toothed cats roaming freely in proximity to their livestock, looking for an evening meal.

Other species are less objectionable on the grounds of how much they might annoy their human neighbors. The mammoth, for example, may be one of the least potentially annoying species on the list of candidate species for de-extinction. The most appropriate habitat for the mammoth is, after all, the Arctic, where human populations are sufficiently small and isolated that the mammoth is not likely to get in their way.

Indeed, Dr. Sergey Zimov, the director of the Northeast Science Station of the Russian Academy of Science in Cherskii, Russia, is intent on recreating the habitat of the mammoth so that it has somewhere to live once its de-extinction is successful. His Pleistocene Park is a nature reserve on the Kolyma River, south of the science station in Cherskii, in northeastern Siberia. Pleistocene Park is in the most northerly part of the habitat that was once the mammoth steppe—the rich tundra grassland that supported mammoths and other grazing herbivores during the Pleistocene ice ages (plate 2). Zimov has already introduced horses from the Urals, bison from eastern Europe, and four different species of deer into Pleistocene Park, and these populations are healthy and self-sustaining. Recently, Zimov decided to expand his operation and establish a second Pleistocene Park in a more southerly location, where the less harsh climate is more amenable to supporting large numbers of herbivores. This second park, which he calls Southern Pleistocene Park, is located in the Tula region, approximately 250 kilometers south of Moscow. Over time, Zimov plans to introduce bison, auroch, horses, wolves, and large cats into Southern Pleistocene Park, which—unlike the Pleistocene Park in northeastern Siberia—is easily accessible by car from Moscow. Both of these locations would likely provide appropriate mammoth habitat, re-creating communities that existed more than ten thousand years ago that should be able to persist without bothering or being bothered by humans.

Will It Be Possible to Learn the Genome Sequence?

With this question, we transition from the big picture to the finer grains of de-extinction. In other words, we now ask whether de-extinction is *technically* possible, or whether it will become so in the foreseeable future. Each of these topics will be covered in detail in the next chapters and so are touched on only briefly below.

The first technical step in de-extinction is to learn the genome sequence of the extinct species. Well, not just the genome sequence. Really, we want to know what the key genetic differences are between the extinct species and any closely related living species. I'll explain what that means in more detail later, but for now let's simply ask: "Can we sequence all of the nucleotides in the genome of this extinct species and then piece those nucleotides together to learn what the sequence of that genome is?"

First, some vocabulary. Genomes are big places, but the molecules that make up genomes are tiny (figure 5). Genomes are made up of chromosomes, which in turn are made up of long strands of *nucleotides*—the building blocks of DNA. Nucleotides each contain a nitrogen base, a five-carbon sugar, and a phos-

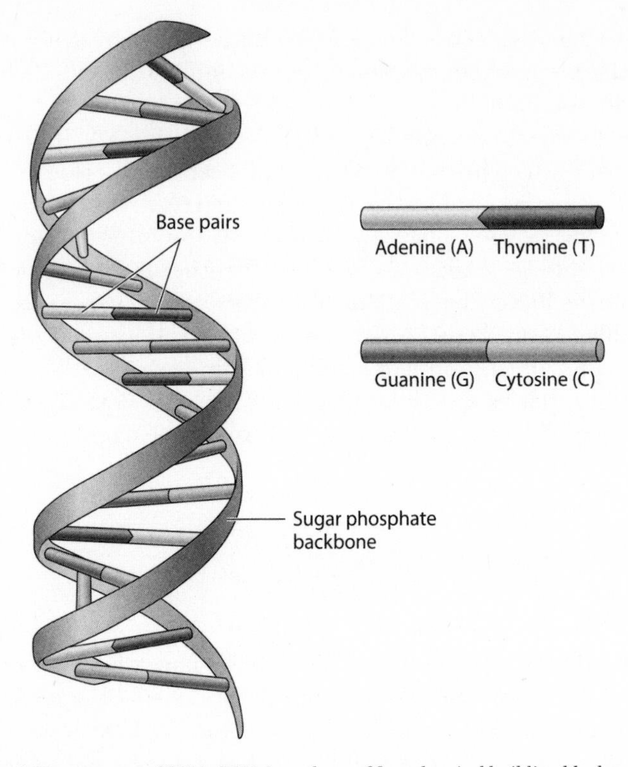

Figure 5. The structure of DNA. DNA is made up of four chemical building blocks called nucleotide bases: adenine (A), cytosine (C), guanine (G), and thymine (T). DNA exists in a winding, two-stranded "double helix" structure, which is formed because the nucleotide bases pair up with each other, creating a ladder-like structure and connecting the two strands. The order of the nucleotide bases, which is also known as the DNA sequence, contains the information necessary to build and maintain an organism.

phate group. DNA genomes contain four different nucleotides, each with a different nitrogen base: adenine (A), guanine (G), cytosine (C), and thymine (T). Nucleotides are strung together along a sugar phosphate backbone to make up *nucleic acids* like deoxyribonucleic acid, or DNA. Genomic DNA is double-stranded, which means that when it is in a stable state, the nucleotide on one strand is bound to a complementary nucleotide on the other strand. A nucleotide that is paired to its complementary nucleotide is called a *base-pair*. Genome sizes are usually reported in base-pairs, which will be half the number of nucleotides.

Genomes vary considerably in the number of base-pairs they contain and in the number of chromosomes among which these base-pairs are distributed. The human genome comprises around 3.2 billion base-pairs, which are found on 23 chromosomes. The loblolly pine genome comprises 22.2 billion base-pairs but only 12 chromosomes. The common carp genome contains 1.7 billion base-pairs distributed among 100 chromosomes. The huge variation in plant and animal genomes is linked neither to the complexity of the organism nor to the number of genes that are encoded by those genomes.

Chromosomes are too long to sequence all at once using existing sequencing technologies. When scientists sequence DNA, they therefore begin by shearing the chromosomes into smaller fragments. These fragments are double-stranded, so their length is also reported as a number of base-pairs. Depending on the sequencing technology to be used, these fragments will vary from fewer than 100 base-pairs long to several thousand base-pairs long. After the DNA has been sheared and sequenced, the fragments are reassembled into chromosomes. To summarize the process of sequencing a genome: First, cut it up. Then, put it back together again.

Now that some of the jargon has been demystified, let's outline the steps of sequencing and assembling the genome of an extinct species. First, we collect remains from the species that we plan to bring back to life—bones, teeth, skin, hair, whatever we can find. Then, we extract and collect as much DNA as we can from those remains. Next, we sequence the DNA. Finally, we take that DNA and carefully assemble the tiny pieces together to make bigger and bigger pieces, and eventually chromosomes.

If you were paying attention, you'll have noted that we skipped the step in which we shear the DNA into smaller fragments. When working with ancient DNA, we don't need this step. The DNA comes pre-sheared. In fact, *over-sheared* is a better way to refer to it. Over-shearing is bad: the shorter a fragment of DNA is, the harder it is to figure out where it goes in the genome.

There is more. These short DNA fragments are also in pretty bad shape. Thanks to chemicals and other biomolecules in the environment, individual nucleotides can become broken or damaged in a way that changes their molecular structure. Molecules with altered structures are read incorrectly during the sequencing process, resulting in mistakes in the genome sequence. The rate of DNA decay is slower in some environments (for example, in the Arctic, where the mammoth lived) than in others (for example, in the tropics, where the dodo lived). This means that species whose native ranges did not include regions of the world where remains are likely to be preserved are probably not ideal for de-extinction.

Finally, we have to deal with what we call *contamination*. Contamination in the broadest sense refers to any DNA that is co-extracted from the bone or other tissue that does *not* come from the organism whose genome we are trying to sequence. It might be DNA from microorganisms that colonized the bone after it was buried in the ground, or from plants whose roots grew around the bone while it was in the ground. It might also be DNA that was introduced into the bone as it was being excavated or handled in the lab. A bone might contain an enormous amount of preserved DNA only a very tiny fraction of which is of interest to us.

Professor Svante Pääbo leads a research group at the Max Planck Institute for Evolutionary Anthropology in Leipzig, Germany, and he and his research group have recently sequenced and assembled the Neandertal genome. His group is very interested in understanding what it means to be human. One way to approach this question is to compare the human genome with closely related genomes of great apes and ask what genetic changes have happened within our genome sequence since we diverged from our common ancestor with other great apes. Our closest living relative is the chimpanzee. The human and chimpanzee genomes are around 98–99 percent identical, which means that what distinguishes us from chimpanzees probably has to do with the other 2 percent. But 2 percent of a 3.2 billion base-pair genome is still *a lot* of DNA to sort through. Neandertals are much more closely related to humans than are chimpanzees. By sequencing the Neandertal genome, Pääbo is able to focus more narrowly on those genetic changes that are unique to our species.

The first complete Neandertal genome that Pääbo's team published was a combination of DNA data that was sequenced from three different Neandertal bones. Each bone contained less than 5 percent Neandertal DNA, with the remaining 95 percent or more comprising mostly environmental DNA—soil microbes and their pathogens, plants, and the like. Of the Neandertal DNA sequences that were recovered from these bones, the average fragment length was forty-seven base-pairs. The human genome contains 3.2 billion base-pairs, so this is a bit like having a puzzle that can only be solved by correctly assembling 68 million puzzle pieces. Of course, thanks to damage and contamination, what they actually had was far more pieces than they needed, some of which were from the same puzzle but cut in a different way and some of which actually belonged to a different puzzle.

To help them to assemble the Neandertal genome, Pääbo's team used the human genome, which was already sequenced and assembled, as a guide. To extend the puzzle analogy, if the forty-seven base-pair fragments of Neandertal DNA were the puzzle pieces, the human genome was the picture on the box top. Only that picture (because it was of a human and not a Neandertal) was not exactly the same as what the puzzle would look like when it was finished. Not identical, but close—maybe the picture was a different color, or maybe part of it was covered by a text box stating "contains very small parts."

Assembling the Neandertal genome was not an easy task. However, it was much easier than assembling many other paleogenomes will be. First, the human genome is the best-resolved genome of any species to date, so the picture on the puzzle box top was nearly complete. The number and diversity of sequenced genomes is growing, but most species genomes are still only partially sequenced and assembled. Second, humans and Neandertals have a common ancestor within the last million years, probably closer to half a million years ago. This means that there hasn't been much time for a huge number of differences to evolve between humans and Neandertals. The picture on the box top pretty closely reflected what the final puzzle would look like.

Not so for many species. In fact, the more evolutionary time that has passed between the divergence between the extinct species and the living species that would be used as a reference, the harder assembling the genome will be. At some point, the picture on the box goes from a slightly discolored version of the end product to something that you rescued from your dog's mouth and then tried to piece together using your imagination and some sticky tape, to something that a herd of mammoths trampled while escaping a pride of cave lions. In the rain.

If there are no remains that contain recoverable DNA, then the species is not a candidate for de-extinction. If there are remains with recoverable DNA, but the species has no close relatives, assembling the genome from that DNA will be challenging—maybe very challenging. Critically, however, it is not impossible to assemble at least large chunks of the genome, even when the DNA that is preserved is in terrible condition.

Is There a Way to Transform the Genome Sequence into a Living Organism?

If we've made it to the step in which we're considering how to create a living organism, we presumably have been able to generate a genome sequence (or a partial genome sequence), even if it might have been tough to do so. Now we have to transform that string of letters into a living thing. How?

There is no clear path from genome to living thing that can be followed for every organism. Some genomes, such as those sequenced from bacterial or viral organisms, are likely to require very little push to start behaving like they are alive. Other genomes are nowhere near becoming a living thing.

Two paths are generally considered when contemplating de-extinction. The first is to do what most people are referring to when they talk about *cloning*. To clone Dolly the sheep in 1996, scientists at the Roslin Institute, which is part of the University of Edinburgh in Scotland, removed a small piece of mammary tissue that contained living cells from an adult ewe and used the DNA in these cells to create an identical copy of the adult ewe. This process is called *somatic cell nuclear transfer*, or, more simply, nuclear transfer. I'll explain how it works later, but for now it is sufficient to know that this is not likely to be the process used to bring many extinct species back to life. Cloning by nuclear transfer, unfortunately, requires intact cells. Unless tissue was taken from a living individual prior to the species' extinction, nuclear transfer will not work. If we're dealing with a species whose genome we have had to sequence and assemble, then we need a different approach.

The other path to creating a living organism is eerily reminiscent of the movie *Jurassic Park*. As is likely to be true in real-life de-extinction projects, Jurassic Park scientists were only able to recover parts of the dinosaur genome—in the case of the movie, from the mosquito blood that was preserved in amber. When they came across holes in the dinosaur genome, they used frog DNA to complete the sequence. Unfortunately, they weren't able to know beforehand which bits of DNA were important to making a dinosaur look and act like a dinosaur, and which bits were just junk. We can only assume that these fictional scientists were hoping that the holes that they were filling in were mostly in the junk-containing regions of the dinosaur genome. But, of course, they were wrong, and some of that frog DNA let the unextinct dinos switch sexes miraculously, leading to disaster and $400 million in box office earnings.

In real-life de-extinction science, the plan is to know which parts of the genome are important in making the extinct species look and act the way it did. We would then find the corresponding parts of the genome of a close living relative, cut out these important sequences, and replace them with the extinct species' version.

Of course, this is all easier said than done.

Let's say we want to bring a mammoth back to life by editing an elephant genome to look like a mammoth genome. First, we have to learn what all the differences between the elephant genome and the mammoth genome are. Then, because making all the changes might be too much to accomplish (at least in the first de-extinctions), we could narrow down which changes to make by deciding which of the differences are important. We might learn, for example, that mammoths have a different copy of a gene called Ucp1—mitochondrial brown fat uncoupling protein 1—than elephants do. Experiments with mice have shown that Ucp1 is involved with thermoregulation. Since mammoths lived in very cold places and elephants do not, we might hypothesize that the mammoth version of this gene helped mammoths to stay warm. Our goal is to turn an elephant into an animal that can survive in cold places, and converting this gene from the elephant version to the mammoth version would help to achieve that goal. So, we construct a molecular tool that can go into an elephant cell, find the spot in the genome that codes for the Ucp1 gene, and edit that gene so that it looks like the mammoth version.

To make the complete mammoth genome, all we have to do is repeat this for every important difference between mammoths and elephants.

Next, we take the cell with the edited genome and inject it into an egg cell that has had its nucleus removed. That cell begins to divide and develop into an embryo, following the familiar path of cloning by nuclear transfer. We then place that embryo into the uterus of a surrogate mom, where it continues to develop and is eventually born.

That last step, in which one species is developing within the uterus of another species, might sound pretty straightforward. However, it also requires some careful consideration. Imagine a project to resurrect Steller's sea cows. Dugongs, the closest living relative of Steller's sea cows and therefore the most likely surrogate mom, have a thirteen- to fifteen-month gestation period, after which they give birth to a single calf. Newborn dugongs weigh about thirty kilograms and are a bit over a meter long—about one-third to one-half the length of an adult dugong. If the same size ratio applies to Steller's sea cows, a newborn calf will be somewhere in the range of three to six meters long. Longer, at birth, than his surrogate mom.

To get around this, one might design a giant sea cow artificial womb. Or, perhaps, a better solution is to choose a species for de-extinction with more suitable options for surrogacy.

Will It Be Possible to Move the Resulting Living Organism from Captivity to a Natural Habitat?

Although much of this has been covered in the answers to the first four questions, I'd like to raise a few additional points here that should be considered when selecting a species for de-extinction. I discussed above whether appropriate habitat exists and what might happen to that habitat and ecosystem if an un-extinct species were suddenly reintroduced. Here, I'm thinking of the more technical aspects of reintroduction. How behaviorally hard-wired was the species? How much parental care was involved with rearing offspring? Were their behaviors learned, or were they born already knowing how to survive, find food, and find a mate? How social were they? Although I brought this up briefly when discussing the additional complexities of bringing back a species with no close living relatives, these are challenges that, to some extent, will face any de-extinction. The first unextinct individual of any species will necessarily be all alone in this world. If behaviors have to be learned, from whom will they be learned? Interaction with the surrogate mom or surrogate community might replace some of the missing social interactions. However, if behaviors are learned from these interactions, will they be the same behaviors that the extinct species would have exhibited? And is that important?

We know from ongoing conservation work that some species survive and seem to do well in captivity but fail to thrive once they are released into the wild. The reason for failure to survive in the wild differs among species. Sometimes animals bred and raised in captivity are easier prey when released, never having been trained to sense and flee from predators. Sometimes they lack the social structure in the wild that they need to be successful. And sometimes they simply never learn how to find their own food. In all of these situations, the only way the species would survive in the wild is if the wild was not actually the wild but instead a site that is actively managed by people. The economic cost of such hands-on management, which may not be small and is likely to take resources away from other conservation and wildlife management programs, needs to be weighed against what is gained by the de-extinction.

THE ELEPHANT IN THE ROOM IS A MAMMOTH

At the beginning of this chapter, I raised the question of who gets to decide what the first targets of de-extinction should be. When I asked the students in my de-extinction class which candidate they thought should have that privilege, they responded with complete silence. Eventually, they offered the only solution appropriate for a group of Californian students: it should be a group decision. But what group? And, even groups have to have a leader, someone who decides ultimately how the group will respond.

The truth is that, at least in these early stages of de-extinction research, the decisions about which species to bring back are going to be made by the people with the interest, money, and expertise to make it happen. The European team working to bring the bucardo back to life is just as unlikely to refocus its attention on the Tasmanian tiger as the Australian team working on the Lazarus frog is to lead the de-extinction of the Yangtze River dolphin. Unfortunately, money is probably the most important determining factor in whether a de-extinction project gains traction. After the death of the baby bucardo in 2003, the bucardo project went silent due to lack of funds. In 2013, after the fresh bout of attention to the project that followed the TEDx event in Washington, DC, the Hunting Federation of Aragón allocated new funds, and the team restarted their cloning efforts. Money, rather than any of the ideas discussed above, may also decide which species are selected for de-extinction. In their campaign to raise funds for de-extinction efforts more broadly, Ryan Phelan and Stewart Brand of Revive & Restore are targeting potential donors on Martha's Vineyard, a wealthy enclave in Massachusetts just south of Cape Cod, asking residents to consider whether they'd like to see Heath hens roaming the island just as they did during the early twentieth century.

And then there is the mammoth. There may be compelling ecological reasons to bring the mammoth back to life, and I will discuss these later. It is also true that mammoths may face fewer technical hurdles in their de-extinction than other species might face. They lived in cold places and many well-preserved bones can be collected and used for DNA analysis. Their closest living relative is the Asian elephant, from which it diverged around 5–8 million years ago, and elephant moms are probably reasonable surrogates for baby mammoths. There is even a place for resurrected mammoths to go: Pleistocene Park is likely to provide a suitable place for mammoths to live, although the steppe tundra that dominated the landscape during the mammoth's reign is not found anywhere on Earth today. That is not to say that mammoth de-extinction would be without challenges. Elephants reach sexual maturity between ten and eighteen years old and have a nearly two-year gestation, which means that genetic engineering experiments will take a long, long time. Also, elephants are extremely social creatures, and there is no reason to suspect mammoths were not highly social as well. Recreating social contexts into which mammoths can be placed will be key to their survival and an additional challenge to overcome.

What inspired the mammoth de-extinction project was not that it would be easy or hard to accomplish, or that it might be ecologically beneficial to have mammoths roaming around Pleistocene Park (although, as I discuss later, the latter is almost certainly true and has become a motivating force as this research continues). The reason that George Church and his group at Harvard's Wyss Institute selected mammoths, rather than kangaroo rats, as the focal species for developing the genetic engineering technology necessary for de-extinction is that mammoths are mammoths whereas kangaroo rats are, well, rats.

Stewart Brand says that his motivation to bring back the passenger pigeon is that these birds, in cultural terms, are as iconic as bald eagles. He believes that the highest value of resurrected passenger pigeons would be to inspire people to be more aware of and engaged in conservation. He puts it more poetically: "Flocks in memory, and flocks in prospect, can make the heart sing." Of course, passenger pigeons are iconic because they formed ludicrously large flocks, which may be difficult to recreate, sustain, or tolerate.

In addition to the challenges of creating and sustaining enormous flocks, passenger pigeon de-extinction faces more (or at least different) technical challenges than does mammoth de-extinction. A high hurdle for passenger pigeon de-extinction is that it is not currently possible to transfer engineered nuclei into bird eggs. There is also no assembled genome sequence yet available either for the passenger pigeon or for its closest living relative, the band-tailed pigeon, but we're working on that (plate 3). We also don't know the extent to which the passenger pigeon was a social creature. The enormity of their flocks suggests they may have been highly social, but whether they need these large flocks to survive remains unknown. The Bronx Zoo, part of the Wildlife Conservation Society, is creating a habitat for the captive breeding of passenger pigeons, but as for whether they will ever be released into the wild, that's also a big unknown. One benefit of choosing passenger pigeons is their generation time. They reproduce every year, so the research necessary to bring them back to life could proceed at a relatively quick rate.

Both the mammoth de-extinction project and the passenger pigeon de-extinction project will require new technologies in order to succeed. But how close are we to seeing either of these species brought back to life? What are the next steps, now that these two species have been selected for de-extinction? Knowing what to do first is easy. First, we have to find the right specimens and extract their DNA.

CHAPTER 3

FIND A WELL-PRESERVED SPECIMEN

One winter morning a few years ago, I met Mathias Stiller and Tara Fulton— two postdoctoral research fellows working in my lab—in a dark, sub-basement hallway of the physics building on our university campus. The dark hallway was home to our ancient DNA lab, which was a purpose-built facility for extracting DNA from poorly preserved samples. Artificial lights flickered ominously overhead as we shed our coats, bags, and shoes, leaving them in the row of outside lockers. Anything that might carry hitchhiking fragments of DNA from the outside world was strictly forbidden to enter the lab. We unlocked the door and moved into the anteroom. The air reeked of the bleach we routinely use to sterilize the floors, surfaces and walls. We dressed in the traditional getup of an ancient DNA scientist: full-body suit, sterile boots, two layers of sterile gloves, hairnet, facemask, and goggles. When we were ready, meaning that no skin or hair or piece of nonsterile clothing was exposed, we moved from the anteroom into the main part of the lab. Tara carried a smoking container of dry ice. Mathias carried an enormous mallet (sterilized, of course). And I carried a tiny plastic bag.

In the plastic bag was a stunning seventeen-million-year-old piece of amber. We had acquired this treasure from my colleague,

Blair Hedges, who purchased it for precisely the purpose we had in mind. The amber weighed around eight grams, measured five centimeters long and three centimeters tall, and was a centimeter or two thick at the center. Encased within the amber were hundreds of tiny bees that had become trapped in the sticky tree resin millions of years ago and, to our eyes at least, were perfectly preserved.

We proceeded to the back corner of the lab, where we had installed a thick, sterile stone plate, above which hung a bright white light and movable magnifying glass. We removed the amber from the bag and wiped it down with a bleach solution so that DNA from anyone that may have touched it over the years would be destroyed. We then rinsed it twice with ethanol to wash off the bleach and allowed it to dry for a few minutes. As it dried, we waited in silence.

When we were certain that enough time had passed, Mathias picked up the amber with sterile forceps and placed it gently into the container of dry ice. Then, we waited again.

Although amber is fossilized tree resin, it is still somewhat malleable—anyone who has ever touched amber jewelry will know what I mean. Stabbing amber with something sharp might dent it, but amber is nearly impossible to break or chip. We wanted to get this piece of amber really, really cold, so that it would be hard and rigid. Brittle.

After ten very long minutes, Mathias picked the amber out of the dry ice with the forceps and placed it carefully on the stone. I then raised the mallet and smashed the little glimmering piece of geological history over and over again until it shattered into a zillion tiny, shining, sticky chunks. Then, using the magnifying glass, we sorted the amber from the bees (plate 4). This involved a lot of re-freezing, re-smacking, and skilled operation of tweezers. After a few hours, we had one tube of mostly amber and another of mostly bees. We took the tube of bees and stuck it in the freezer. We were done for the day.

The next morning, Mathias began the process of extracting ancient DNA from the amber-preserved bees. Over the years, people working in the field of ancient DNA have developed highly sensitive DNA extraction protocols for situations like this one. If DNA had survived in these bees, there certainly wouldn't be much of it left. Mathias opted for the extraction protocol that had proved most successful in recovering very old DNA. We were giving it our best shot.

When the extraction experiments were complete, it was time to send the results off to be sequenced. And then wait. The sequencing results came back three weeks later. We got nothing.

I was disappointed. How incredible would it be to recover DNA from insects preserved in amber? And by incredible, I mean implausible. Far-fetched. Unbelievable. Mathias, I think, was relieved. We both knew that if we did get a result that suggested that millions-of-years-old DNA was preserved, that result would have taken over our lives. We would have had to spend considerable time and energy first convincing ourselves that it was real, and later convincing our colleagues that we had not made a mistake.

When staring into a piece of amber that contains a preserved biological organism, it is hard to understand why the community of ancient DNA scholars would be so skeptical of DNA recovered from that organism. Insects, frogs, and even a 23-million-year-old lizard have been found in fossilized amber, all in perfect physical condition. Why should their DNA not be preserved to the same extent?

The unfortunate truth is that DNA simply does not survive for millions of years. If we did recover authentic ancient DNA sequences from amber, we would have broken all the rules that we've come to understand about DNA preservation and decay.

WHAT? NO *JURASSIC PARK*?

Millions of years before amber is amber, amber is a substance called copal. Thousands of years before it is copal, it is tree resin. Tree resin is a sticky, amorphous, organic substance that is secreted mainly by conifers—pines, cypresses, cedars, sequoias, for example. The resin serves a variety of purposes. It protects the tree from injuries and infections. It may help to heal major wounds, like broken branches. It is also pretty smelly, which might attract curious insects. As the resin oozes out of the tree, bits of plants, insects, and other small animals get trapped and sometimes covered entirely by the sticky substance. Over millions of years, the volatile organic compounds in the resin evaporate, leaving behind only the nonvolatile compounds that make up the amber and anything encased within the amber.

Key to the remarkable preservation of amber-preserved animals is probably the speed in which they are engulfed by resin. If the animal is completely encapsulated and killed almost instantly, that leaves little time for bacteria from the gut or the environment to colonize and start the decomposition process. The tissues also become rapidly dehydrated, killing the enzymes within that would otherwise break down their DNA.

This idea—that amber might provide an exceptional environment for ultra-long-term DNA preservation—is precisely the rationale used by scientists in the early 1990s when they tried this experiment for the first time. Unlike us, however, these scientists claimed success. They shouted it, in fact, in reports that appeared in the most respected scientific journals.

In the early 1990s, the field of ancient DNA was just gaining traction as a serious scientific endeavor. DNA sequences had been recovered from a 170-year-old quagga (an extinct relative of the zebra), from human mummies that were thousands of years old, and from Neandertals and mammoths that were more than thirty thousand years old. Researchers were only beginning to appreciate what this ancient DNA could reveal.

The first applications of ancient DNA were taxonomic: to identify the living species that are the closest evolutionary relatives of extinct species. We now know, for example, that Asian elephants are more closely related to mammoths than are African elephants and that the closest living relative of the dodo is the ornate and beautiful Nicobar pigeon. Some of the taxonomic results from the analysis of ancient DNA have been surprising. In New Zealand, three different species of giant moa

(*Dinornis*) had been described based on differences in the size of their bones. Ancient DNA isolated from these bones showed that, in fact, only one species of giant moa existed on each island. Size, in this case, had nothing to do with taxonomy; the biggest bones were all from female moa and the smaller bones were from males.

As techniques for isolating ancient DNA improved, the field progressed from asking taxonomic questions to asking more detailed questions about the evolutionary history of populations. DNA sequences could reveal cryptic patterns of local extinctions and long-distance dispersals that were invisible in the fossil record. For example, horses—the same species that humans would eventually domesticate—have been around as a distinct taxonomic lineage for at least a million years. Horses originated in North America and dispersed into Asia across the Bering land bridge, which connected the two continents intermittently during the Pleistocene ice ages. Throughout this period, horses dispersed between North America and Asia several times and in both directions, each time establishing new populations and/or hybridizing with populations that already existed. One might even consider the reestablishment of horses in North America by European colonists as the latest in a long history of local extinctions, dispersals, and recolonizations. Feral horses in North America represent, in essence, an unintentional experiment in rewilding that has been extremely successful.

Ancient DNA can identify genes for traits that don't exist anymore, such as mammoth-specific hemoglobin, which makes red blood cells that excel at carrying oxygen around large bodies when it is very cold outside. Ancient DNA can also reveal precisely which genetic changes differentiate humans from Neandertals. To summarize, ancient DNA has turned out to be a very powerful technique for learning about the evolutionary processes that shaped existing biodiversity.

The research group that was leading discovery in ancient DNA during the late 1980s and early 1990s was Allan Wilson's Extinct DNA Study Group at the University of California at

Berkeley. This group of scientists was pioneering the development of protocols to recover fragments of DNA from the remains of dead organisms and, importantly, to distinguish authentic ancient DNA from contaminant DNA.

The science fiction potential for ancient DNA was very quick to catch on. In fact, Michael Crichton acknowledges the Extinct DNA Study Group as part of his inspiration for *Jurassic Park*. And, not long after the 1990 book, science fiction appeared to become scientific fact: several groups (but not the UC Berkeley group) reported sequencing DNA from stingless bees, honeybees, termites, and wood gnats that were tens of millions of years old and even a 120-million-year-old weevil. All of these sequences were generated by extracting ancient DNA from the bodies of insects preserved in amber.

It was too good to be true. In 2013, a team of scientists from the University of Manchester in England performed an experiment to see whether it is possible to extract DNA from bees preserved in copal. Copal, remember, is the precursor to amber and is not entirely fossilized. Copal is therefore much younger than amber. The Manchester team extracted DNA from two copal pieces that contained bees. One of the pieces was around 10,000 years old, and the other was less than sixty years old. They extracted DNA using the latest sample-preparation and DNA extraction techniques. In the end, however, they got nothing—just as we had gotten nothing from our 17-million-year-old piece of amber. They even got *nothing* from the copal specimen that was less than sixty years old.

This Manchester experiment was the second time that scientists tried to extract ancient DNA from copal-preserved bees. In 1997, a team of researchers from the Natural History Museum of London attempted to repeat—and therefore validate—the fantastical results of the early 1990s. These scientists gathered together a variety of amber and copal pieces from their museum's collection and attempted to extract and sequence ancient insect DNA. They also failed to recover any authentic ancient insect DNA.

The absence of results is always challenging to interpret. It is possible that, if one were to generate more and more sequence data, a result might eventually manifest. However, the weight of evidence suggests that ancient DNA is not preserved in amber. Not much is known about what happens to insects once they become trapped in tree resin. Although they probably become quickly dehydrated, which is good for DNA preservation, other characteristics of amber make it an unlikely source of well-preserved DNA. Amber is permeable to gases and some liquids, for example, which means that the DNA may not be entirely isolated from the forces that destroy DNA over time. Also, fossilized amber might be subjected to very hot or high-pressure conditions over the course of its lifetime, both of which are terrible for DNA survival.

The failure to replicate these early experiments proves that DNA is not preserved in amber. What was it, then, that these researchers had been able to sequence in the early 1990s?

GETTING DNA FROM FOSSILS WHEN NO DNA IS PRESERVED

Insects are the most likely source of the insect DNA that was recovered from ancient amber in the early 1990s. Insects that are alive in the present day, that is.

Although I left this out above, the researchers at the Natural History Museum in London were sometimes able to isolate insect DNA from the amber from their collections. In fact, it is precisely this result that led them to conclude that amber was not a source of ancient DNA. In designing their experiment, they selected some pieces of amber that contained encapsulated insects and other pieces that did not. This provided a control: if the DNA was from the amber-preserved insects, then the amber with no insects should have no insect DNA. Their results did not support this hypothesis. They were equally likely to recover insect DNA from pieces of amber that contained insects as they were to recover insect DNA from pieces with nothing in them. The insect

DNA must have been coming from some source other than the animals preserved within the amber.

This result points to a key challenge of working with ancient DNA. In order to recover DNA from specimens that have very little preserved DNA in them, one needs a very sensitive and powerful method for recovering DNA. But, the more sensitive and powerful the method is, the more likely it is to produce spurious results.

In these experiments, researchers were using a technique called PCR—the polymerase chain reaction—to amplify insect DNA (figure 6). PCR was developed in 1983 by Kary Mullis, who, at the time, was working as a biochemist for a company called Cetus Corporation. DNA-sequencing technologies were making it possible to learn the exact sequence of a fragment of DNA. However, to do so, these technologies required millions of clonal copies of the target fragment. Before PCR, this was achieved by enticing bacteria to capture random fragments of DNA within their genomes. These bacteria were then grown into colonies in which each bacterial cell contained an identical copy of the randomly incorporated DNA fragment: enough copies to sequence. PCR provided a much quicker way to copy DNA and, more importantly, a way to target specific parts of the genome to copy. PCR is now one of the most widely used and essential techniques in molecular biology.

Given its revolutionary implications, PCR is surprisingly simple. To walk through the process, imagine that we want to better understand the genetic differences between domestic and wild chickens. A gene *thyroid-stimulating hormone receptor*, or TSHR, is thought to have played an important role in chicken domestication by making chickens reproduce more quickly. We propose to use PCR to amplify—make copies of—this gene from DNA extracts of both domestic chickens and from the preserved remains of ancient chickens that lived prior to the time that chickens were domesticated. We will then sequence the results of the PCRs to learn the sequence of this gene and determine whether domestic chickens have different versions than do their wild relatives and pre-domestic ancestors.

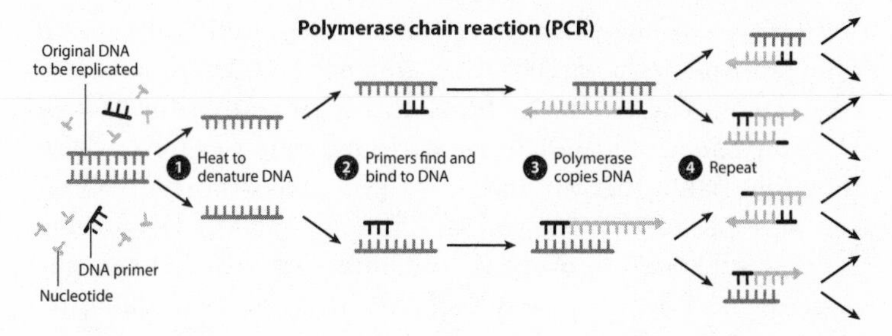

Figure 6. The polymerase chain reaction, or PCR. PCR is a common technique in molecular biology that is used to make billions of copies of a DNA sequence by repeatedly heating and cooling DNA in the presence of a DNA-copying enzyme, free nucleotides to build the copied DNA sequences, and DNA primers, which locate the part of the genome to be copied.

First, we need to have some way to target TSHR. We do this by designing two short DNA probes called *primers*, which will match the sequences of DNA that flank the ends of TSHR. We then make a mixture that contains these primers, the chicken DNA that we already extracted, free nucleotide bases, and a *polymerase*, which is an enzyme whose job it is to copy DNA. Then we can begin the copying process. We heat the mixture to break the hydrogen bonds that hold together the two strands of DNA. When everything is single-stranded, we cool it back down, which causes the strands to come back together. Because the primers are short and there are a lot of them in the mixture, the first thing that happens is that the two primers find precisely the places in the genome that they were designed to match—the regions flanking TSHR—and form double-stranded DNA with that part of the chicken genome. Finally, the polymerase fills in the missing sequence—the TSHR gene—between the primers, using the single-stranded DNA sequence as a template and the free nucleotides to fill in the missing sequence. When this is complete, the number of copies of TSHR has doubled. To make enough copies to sequence, we repeat the process thirty or forty times over the course of a few hours, eventually generating trillions of identical copies of TSHR.

PCR is extremely sensitive. Theoretically, PCR will work if only *one* copy of the target DNA sequence exists in the mixture of extracted DNA. On one hand, this is great news for ancient DNA, where very little DNA is expected to have survived. On the other hand, this is a recipe for potential disaster. If DNA can be PCR-amplified from only one fragment of DNA, then it takes only one fragment of contaminating DNA to ruin the experiment. Given this very high sensitivity to contamination, exceptional results, such as DNA from insects preserved in amber millions of years ago, require exceptional proof of authenticity. At the very least, the result should be able to be replicated. In the chicken experiment I described above, identical experiments were performed in ancient DNA laboratories at Durham University in the United Kingdom and at Uppsala University in Sweden. These identical experiments provided identical sequencing results from ancient chicken remains and, therefore, confirmed that the results were real and not due to contamination.

The main source of contamination in ancient DNA research is DNA from organisms that are alive in the present day. DNA is everywhere. It is on the glassware that is used in the lab. It is in the reagents and solutions that are used to extract DNA. It is on the laboratory benches and the floors and walls and ceilings. It is floating in the air in the labs and hallways. Even more problematically, this contaminating modern DNA is in great physical and chemical condition. Whereas pieces of ancient DNA tend to be broken into very small fragments, mostly of fewer than 100 base-pairs strung together (think "cat," "ant," "bug"), DNA from living organisms can be millions of base-pairs long (think "supercalifragilisticexpialidocious"). Ancient DNA is also broken. Ancient DNA fragments are often missing bases or have bases that are chemically damaged ("cyt," "^nt," "bg"). The polymerase enzymes used in PCR have trouble reading through these damaged sites and end up making mistakes when they copy the sequence ("cut," "int," "bog"). Further complicating things, ancient DNA fragments are frequently chemically linked to other pieces of DNA that are present in the DNA extract, forming knotted molecular structures that the polymerase doesn't recognize as DNA. Because of these problems, the polymerase will preferentially find and make copies of clean, undamaged, freely floating, unbroken, contaminating DNA rather than broken, chemically linked, damaged, ancient DNA. In fact, a single intact fragment of DNA from a living organism can potentially outcompete many hundreds of damaged fragments of ancient DNA during PCR, leading to what looks like, but is not, a real sequence of ancient DNA isolated from a piece of amber. Or from a mammoth bone.

Contamination is not just an idle threat. Contamination comes in many forms, and has played an important role in shaping ancient DNA research. The first and only DNA sequences reported from dinosaurs were (no shock here) contaminants. In fact, many of them were human DNA sequences. Because no one believed dinosaurs were more closely related to mammals than they were to birds or reptiles, and almost nobody believed DNA could be preserved in these very old dinosaur fossils (which are, after all, rocks rather than bones), this result was easily recognized as contamination and rejected.

Contamination is often sneakier, however, and this is when it is most dangerous. DNA from modern pigeons (rock doves, the kind that eat leftover fast food and discarded cigarette butts in city centers around the world) somehow contaminated my very first ancient DNA research project, which was to sequence mitochondrial DNA—a type of DNA that is inherited only down the maternal line (figure 7)—from the dodo. The dodo is, as I mentioned, a pigeon, and I was lucky to have spotted the contamination before writing up my conclusions. In this case, spotting the contamination was pretty straightforward. Whereas most of my experiments failed to produce any DNA at all, one particular experiment produced a huge amount of very high quality DNA. This was a dead giveaway that it wasn't real. I'm still not sure where the contamination came from, but, after that, I started leaving my shoes outside of the ancient DNA lab rather than just covering them up.

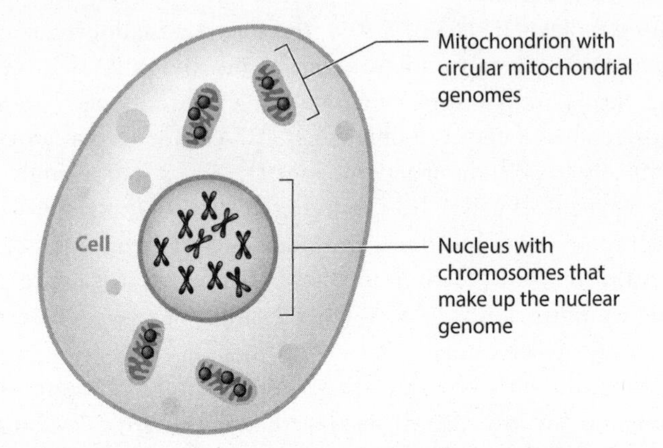

Figure 7. Two sources of DNA in our cells. In humans and other eukaryotic animals, each cell contains two types of genome. The nuclear genome, which includes both the autosomes and sex chromosomes, is found in the cell's nucleus. The mitochondrial genome is found in the mitochondria, which are organelles in the cell cytoplasm. In most eukaryotic animals, mitochondria are inherited only along the maternal line.

In my experience and that of many of my friends and colleagues, there are particular contaminants that pop up from time to time no matter how clean the lab is. DNA sequences of domestic animals and house mice are pretty commonly observed contaminants. This is probably because most of our experiments are designed to amplify DNA from mammals, which, of course, these animals are. Contamination is something we have learned to live with, to expect, and to look for. Because of contamination, we as ancient DNA scientists have developed a healthy wariness of our own data and have set high standards of proof for the authenticity of our results.

This hopefully goes some way to explain the elaborate outfits we wear every time we enter the ancient DNA lab. We are not protecting ourselves from what genetic terrors might be preserved in fossils. Instead, we are protecting whatever DNA might actually be preserved in these fossils from ourselves.

Of course, no matter how careful we are not to contaminate our mammoth bones with our own or any other sources of high-quality DNA, we will likely never find a bone that contains *only* mammoth DNA. In fact, most of the DNA recovered from a randomly selected mammoth bone will be microbial. Which brings me to our next challenge.

THE SURPRISING DIVERSITY OF DNA IN FOSSILS

Let us assume we find a mammoth bone in Siberia and we want to extract DNA from that bone so that we can sequence the mammoth's genome. First, we have to protect the bone from contamination. That means we don't ever touch the bone with our bare hands, because DNA on our hands will get on the surface of the bone, and some of that will be absorbed into the surface layers. We also don't breathe on the bone, stick it in a bag that hasn't been sterilized, or allow it to touch other bones. So, we wear gloves and facemasks and hairnets, and we store every sample separately. When we remove a chunk from the sample to take back to the lab (plate 5), we use sterile cutting instruments, work on sterile cutting surfaces, and clean everything with bleach between samples.

When we return to the lab from the field, we do not remove the sample from the sterile bag unless we are in the ancient DNA lab. There, while wearing our sterile and attractive ancient DNA outfits, we smash the bone into powder using sterile smashing equipment and perform DNA extraction using sterile solutions and sterile lab equipment. After we finish the DNA extraction, we have reduced a chunk of mammoth bone to the contents of a tiny, clear tube: an even tinier amount of liquid that looks indistinguishable from water. In that liquid, we should have mammoth DNA.

And bacterial DNA.

And fungal DNA.

And insect and plant and mouse and dog and human and other DNA.

These non-mammoth DNA sequences are, however, not contaminants. More accurately, they are not contaminants in the same sense that *my* DNA in that sample would be considered a contaminant. The non-mammoth DNA fragments in our DNA extract most likely got into the bone before it was excavated—sometime between when the mammoth died and when we dug up its bone. Bacteria living in the soil, fungi, insects, and plants are all organisms that colonize or grow around bone while it is decaying. Water percolating through the soil will also carry DNA, and this DNA will get into the bone. Even urine carries DNA. A few years ago we showed that sheep DNA can be recovered easily from the same layers of New Zealand soil in which moa DNA is abundant, even though sheep were not introduced into New Zealand until hundreds of years after moa went extinct. A lot of sheep live in New Zealand today. A lot of sheep produce a lot of sheep urine, which leeches through the soil to the deeper layers, commingling with the moa DNA.

Some mammoth bones will have large proportions of mammoth DNA in them relative to the amount of microbial and other exogenous sources of DNA. These are the bones that we prefer to sequence. Unfortunately, it is very difficult to know the ratio of mammoth DNA to other DNA in a sample without going ahead and doing the experiment: extract DNA, sequence it, and see what you get.

Luckily, there are a few general rules about DNA preservation that can be used to guide sample selection. First, cold environments encourage DNA preservation. The chemical processes involved in DNA decay operate more slowly at lower temperatures. Good places to look for bones with well-preserved DNA include the frozen soils (permafrost) of the Arctic and high-altitude caves. Tropical islands are terrible places for DNA preservation, which is bad news for enthusiasts of resurrected dodos (although not all dodos died on Mauritius; some were transported alive to Europe and many of these remains can be found in existing museum collections). Second, ultraviolet light damages DNA. Ultraviolet light causes the same damage to DNA both during life and after death, only dead organisms don't have the DNA repair mechanisms that keep us from getting terrible skin cancers every time we step into the sunshine. This again points to caves as ideal sources of well-preserved remains, and suggests that remains that are rapidly buried are likely to be better preserved than remains that sit exposed on the surface for many months or years. Third, water is particularly damaging to DNA. Rapid postmortem desiccation and preservation in dry or frozen conditions promotes long-term survival of DNA. Ancient DNA has been recovered from naturally mummified remains of humans, steppe bison, mammoths, and other species. Finally, different tissue types tend to be more or less susceptible to damage and decay. Bone, for example, appears to be a better source of intact DNA than is soft tissue, which perhaps has something to do with the structure of bone matrix or with the bone cells themselves. Hair is another excellent source of well-preserved DNA, as the hydrophobic exterior of the hair shaft limits the amount of water and microbes that might enter the hair and degrade the DNA.

THE TEMPORAL LIMITS OF DNA SURVIVAL

The laws of physics and biochemistry tell us that DNA does not survive forever, even in the best environments for preservation. With that in mind, knowing the age of a sample that we're considering for a genome-sequencing project is useful in predicting how successful that project will be. While there is no strict rule that states a precise age beyond which DNA will not survive, biochemical modeling suggests an upper limit close to 100,000 years at moderate ambient temperatures. In practice, however, how old we can go varies considerably, and depends on where in the world the sample comes from, what element (hair, tooth, bone, mummified tissue, eggshell) is preserved, and what the preservation history of the sample has been. In warm environments where samples are immersed in water and exposed to UV light, every bit of useful DNA might be destroyed in less than a year. In the Arctic, if a sample is de-fleshed and immediately frozen and then remains underground and frozen from the time of burial to the time of excavation, the DNA within that sample may survive for many hundreds of thousands of years.

It is important to clarify what I mean by "useful" DNA. DNA does not exist as a beautifully preserved and informative molecule on one day and then dissolve into nonexistence overnight on its expiration date. The process of DNA decay includes both the accumulation of chemical damage and the gradual breaking down of the long strands of DNA into smaller and smaller fragments. Once the surviving fragments are shorter than around twenty-five or thirty base-pairs long, they are too short to map to a unique location in the genome and would therefore no longer be useful for genetic research. Fragments of DNA that are one or two base-pairs long may survive for an extremely long time even in very poor environments for preservation, but recovering these would not help us piece together the genome of an extinct species.

I was recently involved in a large international collaboration to sequence the complete genome of an ancient horse—the same kind of horse that runs the Kentucky Derby today, but a very old one. The bone we used was recovered from permafrost soil in the Canadian Arctic. When we found the bone, we knew that it was old. Very, very old. And we were very excited.

In ancient DNA research, it is crucial to know how old the bones are. Knowing the age of each bone provides a way to correlate changes in population size and genetic diversity with environmental changes. For example, horses went locally extinct in North America around 12,000 years ago. As I discussed in chapter 1, the two competing hypotheses to explain horse extinction are that horses fared poorly through the peak of the last ice age around 20,000 years ago or that they were overhunted by humans that arrived in North America after around 14,000 years ago. Knowing that horses were gone by 12,000 years ago is not the same as knowing *why* horses disappeared. To distinguish between these hypotheses, we need to know *when* horse populations began to decline. And, to learn this, we have to know the age of each bone.

There are several ways to learn the age of a bone, fossil, or archaeological artifact. In some environmental contexts, such as caves or archaeological sites, these items might be found in a clearly defined layer or stratum within which other items whose age is known are also found. There might be a concentration of fossils that are only found together during a particular time interval, or examples of a prehistoric technology that is only used during a specific period of prehistory. Unfortunately, such layers are not common in permafrost, where most of our horse bones are found.

The age of most permafrost-preserved bones is estimated using a process called radiocarbon dating. Radiocarbon dating measures the ratio of two isotopes of carbon—carbon-14 and carbon-12—in organismal remains and uses this ratio to estimate how long ago the organism died. Carbon-14 is a radioactive isotope of carbon that is created in the atmosphere when cosmic rays collide with nitrogen. Carbon-12 is the normal carbon isotope. Both forms of carbon combine with oxygen to form carbon dioxide, which plants absorb through photosynthesis. Animals then eat the plants, and the carbon from those plants is incorporated into their bones. At any point in time, the ratio of the two forms of carbon is the same in the atmosphere as it is in the organisms living in that atmosphere. Carbon-14 is radioactive and decays at a predictable rate with a half-life of 5,700 years. Because organisms stop taking up carbon after death, we can calculate how long it has been since the organism died based on how much carbon-14 is left in its remains.

Radiocarbon dating is a powerful and pleasantly precise way to estimate the age of permafrost bones. However, the amount of carbon-14 in the atmosphere is very, very small relative to the amount of carbon-12—only about one part per trillion of the carbon in the atmosphere is carbon-14—and the half life of carbon-14 is very short. After around 40,000 years or so, too little carbon-14 will remain in an organism to measure accurately. Radiocarbon dating is useful, therefore, only over this very brief, recent time interval.

Fortunately, there is another way to estimate the age of a permafrost-preserved bone. When volcanoes erupt, they send out a wide fan of very fine dust, which is often referred to as volcanic ash or tephra. The tephra produced by each eruption is unique in its geochemical composition. And, as it turns out, geochemists have developed several ways to learn the ages of these volcanic eruptions. These methods are based on the premise that high heat "resets" the age of the minerals, so that different properties of the minerals can then be measured to estimate when the eruption occurred.

Volcanic tephra is deposited across wide swaths of Alaska and the Yukon Territory and marks eruptions of volcanoes as far to the west as the Aleutian Islands and Alaskan peninsula. When the dust settles, a blanket of white forms on top of the permafrost. As time goes forward, permafrost sediments pile up on top of the layer of volcanic ash, which now clearly delineates fossils buried before the eruption, beneath the tephra, from those buried after the eruption, in permafrost above the tephra. This method is not as precise as radiocarbon dating, but it does provide a means to approximate the ages of bones that are too old to be dated using radiocarbon. This is the method we used to learn the age of our very old horse bone.

HOW OLD IS TOO OLD?

My favorite place to do fieldwork is the Klondike gold-mining region just outside of Dawson City, in Canada's Yukon Territory. Gold mining, it turns out, is great for ice age paleontology. Most gold miners in the Klondike use a process called placer mining (plate 6). In placer mining, water from the spring snowmelt is collected into holding ponds. After the sun thaws any exposed permafrost, the water is pumped to the active mining site and blasted against the thawed mud. This washes away anything that is not solid ice. The mining then stops for a little while as the warm sun melts away the next layer of frozen mud. Then, the water is turned back on and the freshly melted mud is washed away. This process is repeated until the permafrost is gone and only the gold-bearing gravels remain.

Much to the bemusement of the miners, we are not particularly interested in the gold. We are, however, very interested in the thousands of bones that are unearthed as the permafrost is washed away (plates 7–9). In the Klondike, around 80 percent of these are bones from the extinct steppe bison, about 10 percent are horses, and the rest are mainly mammoths, bears, lions, caribou, wolves, and muskoxen. Crucially, placer mining is slow and methodical, which means that many of these bones can be picked out of the permafrost while they're *still frozen*. These bones are impeccably preserved.

We found the really old horse bone in a gold mine near Thistle Creek. The site was special, even among Klondike gold mines. A few years earlier, a team of geologists led by Duane Froese from the University of Alberta discovered that the permafrost near Thistle Creek was very old. In fact, it was the oldest permafrost ever discovered. They knew this because they found a volcanic ash layer called the Gold Run tephra associated with the permafrost mud. The Gold Run tephra was deposited across the central Yukon around 700,000 years ago. So, when we found out that horse bones were preserved within 700,000-year-old permafrost, we could not wait to see whether they contained any horse DNA.

Duane recovered seven bones, all of them larger than those of present-day domestic horses, from the layer of permafrost that was associated with the Gold Run tephra. He made sure that the bones were kept frozen rather than allowed to thaw as they were transported out of the field and into storage. We took subsamples from two of these horse bones for DNA analysis and, to our surprise and delight, were able to recover DNA from both. I repeat: we were able to recover authentic, ancient horse DNA from two 700,000-year-old bones.

The fragments of DNA recovered from these horse bones are the oldest ancient DNA sequences that have been isolated from a specimen whose age is well constrained. However—extraordinary claims require extraordinary proof. Were our results real? We think so. We were extremely careful to make sure the sample was kept frozen and kept away from other samples or other sources of contaminating DNA. The fragments of DNA that we recovered from the bones were short and very badly damaged, as is expected when working with ancient DNA. Our analyses of the sequence data indicated that the old horse was evolutionarily more ancient than living horses. And the results were repeatable.

We extracted DNA from these horses in my lab at Oxford and my lab at Penn State, and my colleague Ludovic Orlando and his team at the University of Copenhagen extracted DNA from one of these horse bones several times. The results from all of these extractions were consistent with each other, in terms of both the actual sequences recovered and the damage profile of the recovered DNA. Together, these observations support the authenticity of the very ancient horse DNA.

By the time we finished sequencing ancient horse DNA from this bone, we had generated nearly 12 billion fragments of DNA. We took each of these fragments and tried to match them to the genome sequence of a domestic horse, which had been assembled and published a few years earlier. Around 1 percent of our 12 billion fragments matched different parts of the domestic horse genome, indicating that this tiny component of the DNA recovered from this bone was horse DNA. The other 11.9 billion fragments matched sequences of plant DNA, fungal DNA, bacterial DNA, and other environmental DNA. That is a terrible ratio of horse-to-environmental DNA, and yet we still sequenced the genome of this very ancient horse.

Why did DNA survive in this bone for such an exceptionally long time? We can't say for certain. The bone was found in the oldest permafrost soil that is known to exist, and the bone probably never thawed between the time of burial and 700,000 years later, when we pulled it out of the frozen ground. Unless older, permanently frozen soils are discovered or fossils are recovered from older ice cores, this may be the age limit of DNA survival in bones.

Exceptional preservation is not limited to the Arctic. Caves have also been found to preserve DNA for a remarkably long time. Most of the Neandertal bones that have been sequenced, for example, were recovered from caves. Recently, DNA was recovered from 300,000-year-old cave bears and a 400,000-year-old hominin from bones preserved in Spanish caves. Environmental stability is known to promote DNA preservation, and caves tend to be consistent in both ambient temperature and humidity, which perhaps explains these examples of exceptionally long-term preservation.

Environmental stability does not, however, seem to be an absolute requirement. We recently pieced together the complete 16,000-base-pair mitochondrial genome of a 100,000-year-old bison bone that was recovered from an ancient lake site in Colorado. The bone belonged to an extinct species of bison called *Bison latifrons* whose horns spanned an astonishing 2.5 meters in width—five times the width of today's American bison. The bison bone and the DNA within it somehow survived despite undergoing thousands of seasonal shifts between cold winters and hot summers. The DNA that we recovered from the bone was in terrible condition but, remarkably, remained usable. Would we want to use that particular bison bone as the source of genetic material from which to begin the process of resurrecting *Bison latifrons*? Not unless we absolutely had to. Less than 0.1 percent of the DNA in the bone was bison DNA, the average fragment length was in the range of thirty base-pairs long, and the sequences were badly damaged. But if this bone was the only bone that we had access to and we really wanted to bring the giant bison back to life, we could use this bone to sequence its genome. We would only get a little bit of bison DNA out at a time, and it would be very expensive. But we would probably, eventually, get the sequence mostly correct.

Fortunately for the mammoth and the passenger pigeon, we don't have to rely on badly preserved bones with tiny amounts of DNA. Passenger pigeons died out only a century ago, and hundreds of birds are preserved in museum collections around the globe. Well-preserved mammoth remains are even more abundant. If we limit ourselves to the last 40,000 years—which puts us in the range of radiocarbon dating and allows us to know how old the bones we're working with actually are—there are probably thousands, if not hundreds of thousands, of mammoth remains already in museum and university collections across the world. Most of these are from permafrost deposits, including from the Klondike. Many of these have already been subjects of ancient DNA research, even genome-sequencing projects. We need not be limited, however, to samples sitting on a shelf somewhere at room temperature, subject to faster rates of DNA decay. All we need to do to find an extremely well preserved mammoth bone is get on an airplane, and then a helicopter, and then a boat perhaps, and make our way to the Arctic.

CHAPTER 4

CREATE A CLONE

When you are working in the tundra, nobody cares if you sing loudly and out-of-tune as you walk along a meandering river. Nobody laughs at the five layers of clothing you're wearing or mocks the variety of nets you've donned in your latest, ill-fated attempt to limit mosquito access to your flesh. And nobody bats an eye when your battered Mi-8 helicopter makes an unexpected stop in the middle of the Siberian tundra and picks up a French-speaking couple, their five-year-old child, and a large, red cooler.

These are lessons I learned in the summer of 2008, during what I fondly remember as my strangest and least successful bone-hunting season. That summer, we spent several weeks living in a small encampment surrounded by lakes on the low-lying tundra of the Taimyr Peninsula. We were hunting mammoths.

The Taimyr expedition was led by Bernard Buigues, a seasoned and pleasantly eccentric arctic explorer, and there was no reason to suspect it would not go well. For decades, Bernard, as president of CERPOLEX (CERcles POLaires EXpédition), led overland expeditions into Siberia and to the North Pole. These expeditions began from his well-appointed base in Khatanga, a small Russian town situated on the Khatanga River in Krasnoyarsk Krai. By the early 2000s, Bernard's interests had shifted to expeditions of a more scientific variety, and he had formed Mammuthus, an organization attached to CERPOLEX with the stated goal of exploring and celebrating the Arctic and its many treasures. As the name implies, however, Mammuthus was particularly focused on recovering and facilitating scientific investigations of mummified remains of mammoths. The formation of Mammuthus was either opportunistic or timely, as mummies of mammoths and other ice age giants have been popping out of the Siberian permafrost at a surprising rate since the turn of this century.

Upon meeting Bernard, it was impossible not to have complete faith in both his leadership and the success of the expedition. By 2008, Bernard had decades of experience working in the Siberian tundra. He had seemingly endless energy and enthusiasm, a deep knowledge of the logistical challenges of working in Siberia (and how to circumvent these challenges), and a large collection of warm coats. Most importantly, he had a long history of collaboration with the people living in the region, which goes some way to explain why he is so often the first to get access to newly discovered mammoth mummies. By all reasoning, the expedition should have gone well.

It was at Bernard's Siberian home in Khatanga where our adventure began. Khatanga is an unusual place. It is one of the most northerly inhabited places in the world. Although fewer than 3,500 people live there, it has an airport, a hotel, and a natural history museum full of artifacts relating to the region's people and history. Khatanga also has a few restaurants serving locally sourced meats flavored with dill, and several small stores selling US$8 frostbitten carrots, semiautomatic machine guns, and a bizarre variety of flavored chewing gum. The roads and riverbanks are littered with unfamiliar machines, some of which possibly work. And the people live in anything from small wooden houses to large apartment buildings or even shipping containers—the kind that are stacked on container ships for transport across the ocean. Even Bernard's house was partly made of shipping containers strung together and, presumably, well-insulated. After all, at a latitude of 71°N, winters in Khatanga are dark and cold, with an average low of around−35°C and no sun at all for many days in December and January. We were there in July and

August, however, and the temperature was in the agreeable range of 5°–15°C, with sunshine twenty-four hours a day. Of course, there were a few mosquitoes hanging around, sullying the otherwise pleasant atmosphere. A few hundred mosquitoes, that is.

Per milliliter of air.

Our expedition team included Bernard, his wife Sylvie, and their twelve-year-old nephew Pitou; several Russians who worked for Bernard; a French filmmaker and her boyfriend; and a collection of academic scientists with a variety of interests in ice age animals. The most senior scientist in our team was Dan Fisher, a mammoth specialist and professor at the University of Michigan. Dan is a world expert in deducing everything about a mammoth—its sex and reproductive history, its lifestyle, and even how it died—by studying the growth patterns in its tusks. Dan also measures stable isotopes of elements, such as carbon and nitrogen, that are incorporated into the tusk as it grows. These isotopes contain a near-continuous record of changes in the mammoth's diet and the environment in which it lived. We also had Adam Rountrey and David Fox, both of whom had trained under Dan's supervision during earlier stages of their careers. And there were two of us interested in DNA: Ian Barnes, who at the time was a professor at Royal Holloway University of London but whom I knew from my time studying for my PhD at Oxford University, and me.

Dan, David, and Adam were keen to find tusks, and Ian and I were hoping for bones. Tusks are more useful for isotopic studies, but they contain very little DNA. Ian and I were also interested in the entire community of animals that had lived in the Taimyr during the ice ages, so we were not focused strictly on collecting mammoth bones.

For reasons that remain a mystery to me, and despite promises made to Bernard prior to our arrival in Khatanga, the helicopter was not available for a full week after we arrived. And so we waited. To pass the time while we camped out at Bernard's, we explored Khatanga. We tried on various warm coats and mosquito-thwarting bits of gear. We wandered the streets, taunting local dogs and guessing what the purpose of the various machines might be. We practiced setting insect traps and identifying the things we caught. We drilled holes in a few bones from Bernard's collection for the benefit of the film crew and future research projects. While we waited, Bernard arranged and was occupied by meeting after meeting with his team of Russian scientists and logistical experts. These meetings were colorful and exciting: giant maps were rolled out across tables that were too small to hold them, voices were raised, old scientific papers that outlined the geographical limits of past glaciations were consulted, vodka was consumed, and the excursion was planned.

Finally, the helicopter arrived, and it was time for us to get out into the field. We collected our food, fuel, and gear and headed from Bernard's over to the airport. We maneuvered our way through security to the tarmac and came face-to-face with our next mode of transport: a well-loved Mi-8 helicopter. Two very large gas tanks already occupied about 25 percent of its interior. Working around the tanks, we packed in our camping gear, the cameras and lights for filming, two massive inflatable boats and 250 horsepower outboard motors, enough rice and freeze-dried anonymous foodstuff to feed twenty people for six weeks, a giant petroleum tank for cooking, and enough vodka to keep us happy for at least twenty-four hours. The Mi-8 was missing about a third of its windows, presumably to make smoking onboard more pleasant.

After loading was complete, we climbed aboard and settled in along the benches beneath the windows and on top of the gear and gas tanks. Last to board was Pasha, the cook's dog. Pasha was a one-year-old Siberian husky and was communicating his apprehension about joining the expedition by attempting to fuse with the tarmac beneath the boarding stairs. Pasha and I were on the same page about which was a better fate: being swallowed by the tarmac or taking off in the Mi-8. When it became apparent that the tarmac would not absorb him, Pasha bolted. The cook and one of the pilots climbed out, smoked a few cigarettes, caught Pasha, manhandled him about halfway up the stairs, somehow let him escape, caught him again, subdued him sufficiently to get him all the way up the stairs and through the door, and finally we were set. To the sound of a few cheers and Pasha's desperate howls, we lifted off the ground and headed into the tundra.

SOMATIC CELL NUCLEAR TRANSFER

If so many bones are already housed in collections across the globe, why go out into the field to find more? Why deal with broken helicopters, gold mines, twenty-four-hour daylight, and an infinite number of mosquitoes? The answer is simple: the best bones are those that come straight out of the frozen tundra. We want to find bones that have never thawed. These bones will contain the best-preserved cells, and those cells will contain the best-preserved DNA.

We are not the only team of scientists who spend our summers out in the Arctic searching for the remains of ice age animals or hanging out at placer mines, but I like to think that we are among the most realistic. We know, for example, that we are not looking for cells to clone. What scientists know about cloning animals from somatic cells—cells that are neither sperm nor eggs—suggests that cloning works only when they have a cell that contains an intact genome. No such cell has ever been recovered from remains of extinct species recovered from the frozen tundra.

Degradation of the DNA within cells begins immediately after death. Plant and animal cells contain enzymes whose job it is to break down DNA. These enzymes, called *nucleases*, are found in cells, tears, saliva, sweat, and even on the tips of our fingers. Nucleases are critical to us while we're alive. They destroy invading pathogens before they can do any damage. They remove damaged DNA so that our cells can fix what's broken. And, after our cells die, they break down the DNA in these dead cells so that our bodies can more efficiently get rid of them. This means that nucleases are evolved to remain active after a cell is no longer alive, which is bad news for cloning mammoths.

In the lab, we stop nucleases from degrading away the DNA we're trying to isolate either by dropping a fresh sample in a solu-

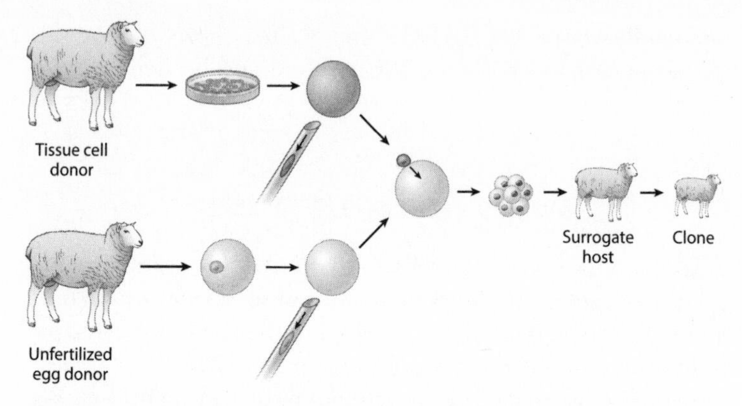

Figure 8. Somatic cell nuclear transfer, or "cloning." A tissue cell (*top left*) and unfertilized egg cell (*bottom left*) are harvested from different individuals. The nuclei are removed, and that from the tissue cell is transferred to the enucleated egg. An electric current is applied, and the egg begins to develop. The embryo is implanted into a surrogate mother and develops into an identical genetic copy of the tissue-cell donor.

tion of chemical inhibitors or by subjecting the sample to rapid freezing. The Arctic is a cold place, but not cold enough to freeze something—in particular something as large as a mammoth—quickly enough to protect its DNA from decay. In addition, all living organisms make nucleases, including the bacteria and fungi that colonize decaying bodies of dead animals. There is little chance, therefore, for any cells to retain completely intact genomes for very long after death. Without an intact genome, there will never be a cloned mammoth. That is to say, there will never be a cloned mammoth via somatic cell nuclear transfer.

Somatic cell nuclear transfer is a dull but appropriate name for the process that brought us, most famously, Dolly the sheep (figure 8). Dolly was cloned in 1996 by scientists at the Roslin Institute in Scotland. Scientists removed the nucleus, which is the part of the cell that contains the genome, from a mammary cell that they harvested from an adult ewe and inserted that nucleus into a prepared egg cell from a different adult ewe. The egg then developed, within the uterus of yet another adult ewe, into a perfectly healthy ewe. Importantly, the ewe that was cloned by nuclear transfer was genetically identical to the ewe that donated the mammary cell and nothing like the surrogate mother or the ewe that donated the egg.

To understand the intricacies of this process, we have to learn some basic facts about the cells that make up living organisms. Our bodies (and the bodies of other living things) are made up of three basic categories of cells: stem cells, germ cells, and somatic cells. Most cells are somatic cells, including skin cells, muscle cells, heart cells, etc. Somatic cells are *diploid*, which means they contain two copies of each chromosome—one from mom and one from dad. Somatic cells also have specialized roles: they might be brain cells, blood cells, or mammary cells like those that were used to create Dolly. Another category of cells is germ cells. Germ cells give rise to gametes, which are sperm and eggs. Sperm and eggs are *haploid*, which means they have only one copy of each chromosome. In normal sexual reproduction, the two haploid gametes fuse upon fertilization to create a diploid zygote, which then develops into an embryo.

In nuclear transfer, the fertilization and fusion step is skipped. Instead, the haploid genome of the egg (a germ cell) is removed in a process called *enucleation*. Then, a diploid nucleus from a somatic cell (in the case of Dolly, a mammary cell) is inserted in its place.

In normal mammalian sexual reproduction, the zygote that is formed at fertilization contains cells that are not at all specialized. These unspecialized cells fall into the third category of cells: stem cells. The stem cells found within early zygotes are called *totipotent* stem cells because they can become any type of cell and are therefore capable of creating an entire living thing. As development proceeds, these cells multiply and begin to *differentiate*, or to take on more specialized roles in the body. Very early in development, totipotent stem cells lose the ability to become every type of cell but are still very unspecialized. These cells are now called *pluripotent* stem cells. Pluripotent stem cells of mammals, for example, can become any type of cell in the body but cannot become placental cells.

Pluripotent stem cells have been of particular scientific interest from a therapeutic perspective. When stem cells divide, they can either make new stem cells or they can become specialized somatic cells. This means that they have the potential to replace cells that have become damaged or diseased. Stem cells are not only found in a developing embryo but are also found in tissues throughout the adult body. Adult stem cells tend to be more specialized than embryonic stem cells but are nonetheless critically important in tissue repair and replenishment. Many medical applications of stem cell therapy make use of adult stem cells. Hematopoietic stem cells, for example, can differentiate into various types of blood cells and are used to treat different forms of blood diseases, including leukemia.

Now back to cloning by nuclear transfer. Somatic cells, unlike stem cells, are highly specialized. Somatic cells cannot differentiate into different types of cells. They are the end of the line, as far as differentiation goes. Somatic cells have a particular job to do, and their cellular machinery is fixed in a way that makes them good at that job. In a somatic cell from a sheep mammary gland, only those proteins required to be a mammary cell are expressed, so only the genes that make those proteins are turned on.

In order for that somatic cell to become an entire living organism, it has to "forget" all of this specialization and de-differentiate. It has to turn back into an embryonic stem cell.

While Dolly is perhaps the most famous animal born via somatic cell nuclear transfer, she was not the first clone to be produced in this way. In the 1950s and '60s, John Gurdon of Oxford University showed that frog eggs would develop into frogs even after their nuclei were removed and replaced with nuclei from somatic cells. Although the mechanism was not well understood at the time, Gurdon's key observation was that the egg somehow triggered de-differentiation of the somatic cells—they forgot what type of cell they were. In 2012, Gurdon shared the Nobel Prize for these discoveries with Shinya Yamanaka of Kyoto University, who later discovered that the same pluripotency (de-differentiation of somatic cells) that was induced by the egg could be induced in vitro, that is, in tissue culture in a lab rather than in an egg, by adding a suite of transcription factors, which are proteins that bind to specific DNA sequences and control what genes are turned on and when. Such cells are known as *induced pluripotent stem cells*, or iPSCs.

Nuclear transfer has been used to clone sheep, cows, goats, deer, cats, dogs, frogs, ferrets, horses, rabbits, pigs, and many others. Cloning animals with specific sought-after traits is also gaining popularity. Commercial services to clone pets and to provide cloned offspring of champion horses are advertised widely on the Internet. The results of such selective cloning have begun to manifest: in late 2013, Show Me, a six-year-old clone of a polo-playing mare named Sage, won the Argentinian Triple Crown, perhaps ushering in a new era of animal breeding for show and sport.

Cloning by nuclear transfer is not particularly efficient, however. Dolly was the only one of 277 embryos created by the Roslin Institute that survived to be born. The first cloned horse to be born, a female named Promotea, was the only one of 841 embryos to fully develop. Snuppy, an Afghan hound cloned by the Korean scientist Hwang Woo-Suk, was one of two puppies born after 1,095 embryos were implanted into 123 different surrogate mothers, and the only puppy to survive for more than a few weeks. In each of these cases, scientists had access to a potentially limitless supply of somatic cells, all harvested from *living* animals.

There are no living mammoths.

SEARCHING FOR A MIRACLE

In the last several decades, sites rich in extremely well preserved frozen bones have been discovered across Siberia, Alaska, and Canada's Yukon Territory. This area, collectively known as Beringia, was an important conduit for movement between Asia and North America during the Pleistocene. Based on the number and variety of bones collected from across Beringia, the area was teeming with megafauna—animals that weigh more than forty-five kilograms—throughout the Pleistocene. The remains of the Beringian megafauna are exposed when the permafrost in which they are buried is disturbed. We disturb the permafrost by building towns, building roads to connect the towns, and looking for gold. Ice age bones are also exposed through natural processes, such as the annual flooding of rivers and lakes after the spring snow melts (plate 10). High- and fast-flowing water rips around river bends, tearing into the frozen dirt along the river edges and washing out any bones or other megafaunal remains that had been frozen within the dirt.

Up in the Taimyr, Bernard selected a site for our base camp that he felt was a prime location for bone hunting, based on the hours he spent consulting maps and conversing with locals. We pitched our tents near the top of a relatively high, large hill within a landscape that was mostly water separated by patches of low-lying, treeless tundra (plates 11–13). Our plan was to walk the perimeters of the many lakes and connecting waterways, keeping our eyes peeled for bones or tusks.

I have spent many summers of my life searching for ice age bones in Beringia. Mostly it's a lot of the same: wandering along rivers and lakes staring into the shallow water, or hanging out at active mining sites and waiting for the hoses to be turned off so that we can scan the freshly thawing surface for ice-age treasures. And nearly every day that I have spent in the field has been wildly productive.

Our first day in the Taimyr was unproductive. We set up our own tents, the cook's tent, and our "rest" tent, which was really just a frame set inside a giant mosquito net that gave us enough space to crowd around a table away from the bloodthirsty onslaught. We inflated the boats and got them ready to go. We set traps to catch fish. We scoped out the edges of the lakes that were closest to us. We ate rice and fish, and celebrated our arrival in the field with a toast. And we found zero bones.

The second day was also not productive. We took the boats out and walked along the edges of lakes that were slightly farther away. We donned chest waders and ventured deeper into the freezing water. We found no bones. We returned to camp, and ate a dinner of rice and fish.

The third day was also unproductive. We split into smaller groups to scout out different nearby lakes, but nobody had any luck. That night, we sat in silence in our mosquito-free enclosure, eating our rice and fish. I'd never been on an expedition for three days and not found a single bone. I think the same was probably true for all of us. The glamour of being on an arctic expedition had pretty much worn off after the first seven thousand mosquito bites, and we'd already finished the vodka. To say the mood was bleak would be an understatement. Here we were, facing at least another several weeks on the tundra, with no idea why there were no bones to be found and no sense of what to do about it.

And then two things happened. First, we heard a rustle outside the enclosure and looked up to see two men who were not part of our expedition team standing there, quietly, with shotguns. Then, the French couple opened their cooler.

RENEWED HOPE AND THE BEASTS OF THE UNDERWORLD

More mummies have been recovered from permafrost deposits in Siberia than from permafrost deposits in North America. This may be because mammoth populations were larger in Siberia or because some aspect of the climate makes preservation of mummified bodies more likely in Siberia than in North America. Whatever the reason, the discovery of a mammoth mummy always causes a stir. For many of the indigenous people of the Siberian tundra, that stir is deeply personal. Some cultures have mythologies that refer to mammoths as beasts of the underworld and caution that touching them will bring bad luck—even death—to the unfortunate discoverer. More widely, though, the stir is one of excited anticipation. A mummified carcass is a special thing—and one for which scientists may be willing to pay a high price.

Some of the mummies that have been recovered from the Siberian permafrost are impeccably preserved, with intact tissues, hair, and internal organs that are clearly visible in CAT scans and autopsies. Oddly, the DNA within even the best preserved mummies tends to be in bad shape compared with the DNA that is preserved in bones. One possible explanation is the difference in the amount of time it takes to freeze the DNA. If body parts are scavenged and the flesh consumed by predators, the de-fleshed bones are likely to be rapidly buried and frozen in permafrost, while mummies would stay warm for far longer. While the mummy was slowly freezing, microbes from the animal's gut and the environment would colonize tissues throughout the body, decomposing the animal from the inside and simultaneously destroying the DNA.

Although the record of DNA preservation in mummies is startlingly poor, we can't seem to separate the remarkable physical preservation of their bodies from the idea that their DNA *must,* somehow, be equally well preserved. With each find, there is renewed enthusiasm that *this* mummy will be the one that defies the odds. *This* is the mummy that will have intact cells with intact nuclei that contain intact genomes. *This* mummy will have the donor cells for cloning by nuclear transfer.

The first I heard of Bernard Buigues was just after one of these remarkable finds. It was October 1999, and a mammoth that no doubt had intact cells and intact nuclei with intact genomes had just been flown across the Siberian tundra.

Whenever there is a spectacular result in the ancient DNA world, my colleagues and I are inundated with calls from journalists looking to be the first to break the story about the imminent resurrection of the mammoth/dinosaur/dodo. On this particular day, I was sitting at my desk in Alan Cooper's ancient DNA lab at the University of Oxford. It was my first month as a PhD student and immigrant to the United Kingdom.

The phone rang, and I answered it. The caller launched into a series of rapid-fire questions, speaking in an accent that was unfamiliar to my American ears. I made out the words "helicopter" "jackhammer" "cryogenics" "tusk" and "Siberia" but did not manage to find a break into which insert a response (such as "Could you please call back when someone who's been at this for more than two weeks is in?"). The journalist then paused and, much more clearly, asked me what my opinion was about whether a hair dryer could ruin the chances of cloning a mammoth.

I was pretty sure I *could* have an opinion about a hair dryer and its role in cloning a mammoth. I was also sure that, since I wanted eventually to be taken seriously as an ancient DNA scientist, I should probably ask for clarification before offering an opinion.

I learned that a team of arctic explorers led by my soon-to-be friend and colleague Bernard Buigues had just unearthed what they believed to be a nearly complete mammoth mummy. In a drastic and dramatic attempt to keep the mammoth cells frozen and therefore intact, they left the slightly decaying corpse in the ground until winter so that the ground was good and frozen. Then, using jackhammers and strong shovels and working in the freezing dark, they cut a 21,000 kilogram block of frozen dirt out of the permafrost and flew it, hanging off the underside of a large helicopter, nearly three hundred kilometers back to Bernard's underground cave in Khatanga, where they planned to slowly and methodically thaw the mammoth carcass out of the ice using a hair dryer.

For good measure, and because it made the pictures and video even more impressive, Bernard (who admits that he was using "creative license" when he did this) stuck the tusks that had been found near the exposed skull into the side of the frozen block of ice before the helicopter took off, so that it looked like there was a complete mammoth inside a frozen box flying across the tundra. They knew that the mammoth carcass in the block of ice was incomplete. They had already removed the head, for example, which had partially thawed and begun to rot. They had also used ground-penetrating radar to try to see beneath the surface, and the results hinted that less than a complete mammoth was preserved within. But they were hopeful.

This mammoth, which was named Jarkov after the local family who discovered it, lived around 23,000 years ago. Jarkov was an adult male mammoth, about three meters tall, that probably died a few years before his fiftieth birthday. The idea that Jarkov could be cloned was floated almost immediately. This idea was embraced in particular by the Discovery Channel, which funded

Jarkov's dramatic extraction from the ground. Larry Agenbroad, a mammoth expert from Northern Arizona University, reported in the team's press release that they had already lined up a lab with expertise in cryogenics and "elephants available."

A year later, the hair-dryer defrost revealed only a small amount of mammoth preserved within the giant block of dirt. Even more disappointingly, what was preserved was mostly bone, with a bit of tissue and some hair. No intact nuclei were discovered, but short fragments of DNA extracted from the hair were used to construct a complete mitochondrial genome and, eventually, part of the mammoth nuclear genome. Jarkov would not be the first cloned mammoth. However, the spectacle of his extraction from the earth and flight across the tundra instilled in the public a sense of just how important a *frozen* mammoth would be for mammoth cloning. It also reinforced the (incorrect) assumption that what we really needed to find was a whole, perfect mummy.

One year before the spectacle of the Jarkov mammoth flying across the tundra, a team of Japanese scientists led by Akira Iritani and Kazufumi Goto founded the Mammoth Creation Project, whose goal was clearly stated in the name. Iritani and Goto were involved with in vitro fertilization research in Japan, and both had made fascinating discoveries about the hardiness of sperm. For example, they learned that sperm taken from cows and pigs and frozen to −20°C could be defrosted and used to fertilize eggs, from which perfectly healthy cows and pigs would develop. Having read about Zimov's Pleistocene Park, they wondered whether frozen mammoth sperm might be key to resurrecting the park's star attraction.

With mammoth sperm on his mind, Iritani set off on a series of Siberian expeditions in search of a frozen bull mammoth. The expeditions were led by Petr Lazarev, a geologist and head of the Mammoth Museum in Yakutsk. If they were successful in finding bull mammoths, Iritani and Goto planned to harvest their sperm and use it to fertilize the eggs of elephants. Because this would result in a hybrid calf and not a cloned mammoth, they intended to use sperm that contained the X chromosome and make only females. Then, when the hybrid females became sexually mature, they would impregnate her with embryos created using her eggs and other mammoth sperm. In this way, Iritani predicted that he would be able to create a creature whose genome would be 88 percent mammoth within only fifty years.

After two summer expeditions in 1997 and 1998, the Mammoth Creation Project had no money left and no mammoth sperm to show for their effort.

Then, in 2002, the Yukagir mammoth was discovered.

A FIRST ATTEMPT

In the autumn of 2002, Vasily Ghorokov was out hunting for tusks along the banks of the Maxunuohka River in Yakutia, northern Siberia. Ghorokov and his sons spotted the tip of what looked like a particularly well-preserved specimen and began to dig. As he reached the base of the tusk, he realized that it was still attached to what turned out to be most of a skull, which was so well preserved that parts of it were covered in skin and hair. Word spread quickly about the new find, and competing groups hurried to find a way to reach the site. Buigues learned of the find via his extensive connections across Siberia. In Yakutsk, the news reached Lazarev at the Mammoth Museum. Lazarev called Iritani and revealed plans to continue the excavation the following autumn. And Iritani decided that at seventy-one, he was too old for yet another Siberian expedition. He would instead send one of his students.

A year later, a team of international scientists arrived at the site of the Yukagir mammoth. Buigues led the team, which included, among others, Iritani's student Hiromi Kato, Petr Lazarev, and Alexei Tikhonov, the scientific secretary of the Russian Mammoth Committee who was based at the Zoological Institute in Saint Petersburg. In this second season of excavation, the team painstakingly recovered the mammoth's left front leg, taking extreme caution to keep it frozen. Like the skull, the leg was impeccably preserved and covered in soft tissue and hair.

Then the trouble began. A rival Japanese team surfaced and offered a hefty reward to anyone who could provide a mammoth that could be a main attraction at the upcoming 2005 World Expo. Export permissions became impossible to obtain. In the end, the leg had to stay. Kato returned to Iritani empty-handed, no closer to a cloned mammoth. Lazarev, doing his part, snagged a bit of the foreleg tissue and carried it personally to Iritani in Japan, but by the time he arrived the flesh had begun to decay.

After one more autumn excavation, this time directed by Naoki Suzuki of Jikei University in Tokyo, the Yukagir mammoth was removed entirely from its tundra grave. Parts of the vertebral column and rib cage were recovered, as was a portion of intestine packed with fecal material. Scientific analysis of these remains revealed that when the Yukagir mammoth died around 22,500 years ago, it was forty-eight years old and weighed somewhere in the range of 3,500 to 4,500 kilograms, which was average for an adult male mammoth. Suzuki would eventually oversee transport of the Yukagir mammoth to Japan, where it would be intensively studied using X-ray computer tomography, providing the first internal anatomical scan of a mammoth without causing any harm to the specimen. While in Japan, the Yukagir mammoth was a centerpiece at the 2005 World Expo in Aichi.

After its stint in Japan, the Yukagir mammoth was flown back to Yakutsk, where it is currently stored in an underground cave in the center of the city, where frozen fish and reindeer and other food is stored (plate 14). A few summers ago, I had an opportunity to see the Yukagir mammoth myself. It sits in the far back corner of the cave, in a compartment of its own. The Yukagir mammoth is as impressive as the hype suggests. However, no intact mammoth cells have been recovered from its body, despite its extremely good state of preservation.

A few years ago, Iritani and his team published a research paper in the *Proceedings of the Japan Academy*, in which they described the first experiment to clone a mammoth using nuclear transfer. Iritani's team extracted cells from the bit of foreleg that Lazarev managed to get out of Russia, including cells from what appeared to be preserved bone marrow. Iritani's team prepared mouse eggs for nuclear transfer by removing the mouse nuclei. They inserted nuclei that they managed to extract from the Yukagir mammoth's cells into the prepared mouse eggs. If the genomes within these mammoth cells were sufficiently intact, the mouse egg would hopefully trigger the mammoth somatic cells to de-differentiate into stem cells, and development would begin.

Nothing happened.

A BETTER MAMMOTH AND A POSSIBLE SOLUTION TO THE PRESERVATION PUZZLE

In 2007, the three sons of Yuri Khundi, a Nenet reindeer herder, discovered a nearly perfectly preserved baby mammoth along the banks of the Yuribey River in northeastern Siberia. Khundi wanted the mammoth, but he wasn't sure how to get it out of the tundra. The Nenets believe that mammoths are bad luck—beasts that wander the darkness of the frozen underworld. Electing not to risk the retribution of the beasts, Khundi and a friend decided to see whether the director of a local museum had any ideas. Sensing something important, the museum director convinced the local authorities that they should help. The entire crew then went back to the Yuribey River. When they got there, there was no baby mammoth.

It turned out that one of Khundi's cousins had heard the story of the baby mammoth by the river and, less concerned with bad luck than with good fortune, decided to go get it himself. Khundi was not happy about the turn of events. He found out that his cousin had been seen heading for a nearby town, so Khundi and his friend followed. When they arrived, they found the mammoth propped against the wall of a store and looking a little bit worse for wear. Khundi's cousin had sold the mammoth to the store's owner in exchange for a year's worth of food and two snowmobiles. Unfortunately for the mammoth, local dogs had been chewing bits off of its extremities whenever the store owner's back was turned.

The story has a happy ending: Khundi managed to reclaim the baby mammoth before much more damage was done, and the mammoth was moved to the Shemanovsky Museum in Salekhard for safekeeping.

The mammoth, a female that was later named Lyuba, was only a month old when she died 42,000 year ago. She was so well preserved that her stomach still contained traces of her mother's milk. About a year after her discovery, researchers including Bernard Buigues, Dan Fisher, Alexei Tikhonov, and Naoki Suzuki performed a marathon three-day autopsy of Lyuba's body in a lab in Saint Petersburg, Russia. They discovered fine mud in her lungs, mouth, and throat, which likely meant she had died of asphyxiation, perhaps while trying to cross a muddy river. They studied her baby tusks, looked for mites in her hair, and learned that, like elephants, mammoth babies ingest their mother's feces to inoculate their digestive systems with the microbes that will break down the plants they eat. And, in an important step for anyone interested in cloning a mammoth, they discovered why Lyuba was so well preserved.

Dan Fisher, one of the members of our expedition team during that unproductive summer on the Taimyr, was key to solving this puzzle. Dan is a soft-spoken man who knows a lot about mammoths. His interest in mammoths, however, is not limited to the animals themselves. He also cares a great deal about how people interacted with mammoths. For example, mammoths were certainly too big to eat in one sitting. One question that Dan seeks to answer is, how did mammoth hunters preserve meat in the absence of modern refrigeration?

While we were in the field, Dan told us about a series of experiments that he had performed near his home in Michigan to see how long meat would remain edible if it were stored in shallow ponds. First, he butchered lamb and venison and anchored the meat to the bottom of shallow ponds in a nature reserve associated with his university. Over a period of two years, he would bring up the meat every now and then and check for decomposition. Then, one day in mid-February of 1993, a colleague gave him a draft horse that had just died of natural causes. This gave

Dan a new idea. Using stone tools that he fashioned himself, mimicking as best he could the technology of the mammoth-hunting indigenous people from the Great Lakes Region, he butchered the horse. It was winter and the ponds were covered in ice. So he chopped a hole through the ice and submerged the horse meat in the cold water. Every two weeks, he brought the meat out and cut off a piece to test for palatability and signs of decay. By June, Dan noted that the meat, while still retaining considerable nutritive value, had developed a sour taste and a strong, sour odor. This was the same strong, sour odor that Dan noticed coming from Lyuba's carcass as they performed their autopsy in Saint Petersburg.

The sour odor was caused by microbes called lactobacilli. Lactobacilli convert lactose and other sugars to lactic acid and are found naturally in the guts of many animals. The buildup of lactic acid in Lyuba had effectively pickled her, preserving her in the permafrost where she was buried and protecting her from decay even after her body was exposed.

Unfortunately, although high acidity might be good for pickling mummies, it is not good for DNA preservation. These mummies may appear to be very well preserved, but the high-acid environments cause considerable cellular damage and destroy naked DNA. That means that, while these mummies might appear—superficially—to be the most likely source of an intact cell suitable for cloning, their remains may actually be the worst place to look for such cells.

Some scientists remain undeterred, however, and the race to clone a mammoth remains in full force. Teams are still out every summer looking for mammoth mummies, hoping that one day an exceptionally preserved mummy will emerge, unpickled, from the Siberian tundra.

UPPED STAKES AND A NEW CONTESTANT

In 2008, Teruhiko Wakayama of the Tiken Centre for Developmental Biology in Kobe, Japan, cloned mice that been frozen at−20°C for sixteen years. This was a huge and important step for the de-extinction effort, for two reasons. First, all of the cells that Wakayama and his team used were dead before they were injected into prepared mouse eggs. That meant that mammoth hunters might not need to find a living cell in order for nuclear transfer to work, because, sometimes, even dead cells contain sufficiently intact genomes for cloning. Second, they discovered that they could increase the chances of success of nuclear transfer by adding a step to the cloning protocol. Their results suggested that some cells, particularly those whose genome might be a little bit broken, may simply need an extra push in order to be fully de-differentiated.

Initially, Wakayama's team followed the standard protocol of nuclear transfer: isolating nuclei from the once-frozen mouse cells and inserting them into prepared mouse eggs. Although not many of the eggs began to develop, a few did, indicating that the egg was able to reset some of the somatic cells. However, none of these went on to become fully developed mice. Instead, the process stalled after a few cell divisions, suggesting that de-differentiation was not completely successful.

Then they had an idea. They repeated the process, but this time they stopped the embryo from developing after only a few rounds of cell division. They then took those cells that had started to develop and used them to create what are called *cell lines*—large colonies of identical cells growing in the lab. Next, they removed nuclei from these growing cells and inserted them into a freshly prepared egg. In this way, the egg had not one but two chances to reprogram these cells into completely differentiated stem cells. To the astonishment of the scientific community, two of the embryos created in this way went on to develop into healthy, adult mice.

It was this experiment that motivated Iritani and his team to try to clone cells from the Yukagir mammoth's leg. Although Iritani's team was unsuccessful (none of the mammoth cells developed to a stage where it was possible to try to make cell lines), he remains undeterred. His team had, after all, managed to isolate a nucleus from a mammoth cell, which was a remarkable feat in itself.

In August 2011, a mammoth thigh bone was found in the Sakha Republic that was so well preserved that it still contained greasy bone marrow. Certain that this was the ticket to cloned mammoths, Iritani used this find as a springboard to reinvigorate his mammoth-cloning plans. That December, Iritani announced that he would clone a mammoth by 2016. His timeline required that (1) they find a perfectly preserved mammoth during the following field season; and (2) they would be able to establish cell lines from that mammoth immediately. Given that elephants have a 600-day gestation period, his plan left no room for error.

Iritani's announcement was embraced by the global media, which delighted in the opportunity to publish yet another round of mammoth-cloning-is-inevitable stories. The most intriguing response, however, came from South Korea, where one more contestant in the race to clone a mammoth was about to emerge.

In March 2012, Hwang Woo-Suk at the Sooam Biotech Research Foundation announced with great fanfare that Sooam had established a new collaboration with the North-Eastern Federal University in Sakha (with which the Mammoth Museum is affiliated, and with which Iritani had been working since 1997), and that *he* was going to clone a mammoth. The announcement went viral, complete with pictures of a smiling Hwang shaking hands with Vasily Vasiliev, vice-rector of North-Eastern Federal University, over official-looking documents. Almost immediately, the *Moscow News* published a clarification of the report. Without identifying its source, the *Moscow News* stated in strong and clear language that while the Russian Academy of Sciences certainly was planning to clone a mammoth, it would do so in collaboration with Iritani and the team from Kinki University and *not* with Hwang.

It is not surprising that reaction to Hwang's involvement in the high-profile cloning project would bring about mixed emotions. I mentioned Hwang earlier in this chapter, with a brief reference to his work to produce the first cloned dog, Snuppy. Hwang is, however, better known for his work in *human* cloning. In the early 2000s, Hwang was leading a research group at Seoul National University that was at the absolute cutting edge in human stem cell research. His group published two major breakthrough papers in 2004 and 2005. The first claimed that they had cloned the first human embryos, and the second indicated that they had made stem cells that were genetically matched to specific people; these were enormous advances for biomedical research. In Korea, Hwang was praised widely as a national hero. And then the walls came crumbling down. In 2006, Hwang retracted both papers after it was revealed that the data had been faked. He lost his job at the university and was stripped of his license to conduct stem cell research. He was also charged with fraud, embezzlement, and bioethics infractions and was eventually found guilty of the two latter counts.

Hwang's trial lasted for three years from 2006 to 2009. During this time, he joined the Sooam Biotech Research Foundation and continued his research, which was now focusing on cloning animals. The first official mention of Sooam's plans to clone a mammoth came in 2012, with the announcement of collaboration with the North-Eastern Federal University. Hwang's interests were already well established by that time, however. During his trial in 2006, Hwang explained why so much of the standard documentation of research expenses was missing from his files: he needed to pay the Russian Mafia for access to the best mammoth carcasses.

In the autumn of 2012, on the heels of their big announcement, Hwang Woo-Suk and his student, Hwang Insung, joined Semyon Gregoriev of the North Eastern Federal University on a three-week expedition up the Yana River to find a mammoth to clone. The trip was being filmed for National Geographic by a London-based documentary maker who intended to tell the story of Sooam's project from start to, well, start. Although the expedition did not succeed in finding a mammoth mummy, reports emerged as soon as they returned from the field that a remarkably well-preserved piece of skin had been found buried in the frozen ground. Most importantly, the skin was said to contain cells with intact nuclei.

A few weeks prior to the trip, the filmmaker contacted me about joining the expedition as the genetics expert. Unfortunately, I had to stay behind (to give birth to my second son), but recommended my friend and colleague, Love Dalén, who runs an ancient DNA lab at the Swedish Museum of Natural History. Love tells a slightly less fantastic rendition of the story than the documentary portrays. In Love's version, the team already knew where to look for a mammoth before the expedition began. Yakutian mammoth hunters had spent the early part of the season looking for tusks along the river. In doing so, they blasted a series of long tunnels into the permafrost along the riverbanks using high-pressure water. At the end of one such tunnel, someone had spotted a perfectly preserved baby mammoth. The mammoth—sans tusks, of course, as these would already have been removed by the first people to encounter the freshly exposed mummy—was still in place, and the plan for the show was to go back and get it. Unfortunately, by the time the expedition team arrived and filming began, late season rains and flooding had caused the tunnel to collapse, leaving the expedition/documentary team desperately searching those tunnels that remained for anything that would suffice for their show. The skin in question was found by Hwang Insung after he maneuvered into one such tunnel despite warnings from the expedition's safety officer that it was dangerous to do so. Hwang found the piece of skin deep within the tunnel, just before getting word from the outside that the tunnel was about to collapse. After a few moments of desperate panic, those who had dared enter the tunnel emerged, narrowly escaping being crushed by several thousands of kilograms of frozen dirt.

Did the bit of mammoth skin that they found have cells that contained intact nuclei? Perhaps. Finding what appears to be cellular structure is not uncommon in permafrost-preserved remains. Will the genome within those cells be sufficiently intact to be cloned? Doubtful. Love was able to take a subsample of the specimen to Stockholm where he extracted and amplified DNA.

He confirmed that the skin was, in fact, from a mammoth. But the longest fragments of DNA that Love could amplify were around 800 nucleotides long. That is a remarkably long fragment for ancient DNA (the average length of fragments from permafrost-preserved specimens is closer to seventy nucleotides long), and it certainly indicates that the specimen is well preserved. Still, 800 nucleotides is a far cry from the length of an intact chromosome.

In the summer of 2013, a new partial mammoth carcass was discovered frozen in a lake on Malolyakovsky Island, part of the New Siberian Islands. The find was absolutely stunning. The part of the mammoth that had been exposed was beginning to rot, but other bits of flesh were so well preserved that they were described as looking like fresh meat. Most intriguingly, a deep red substance suspiciously reminiscent of blood was found in the permafrost beneath the carcass. While most experts (myself included) are highly skeptical that the substance actually is blood—there is no animal with blood capable of staying unfrozen in the conditions in which this sample was recovered—research has been inconclusive so far with regard to what it actually is. The specimen has been kept frozen and is currently being studied in Yakutsk by scientists from around the world.

Is this latest mammoth the "best preserved mammoth in the history of paleontology," as Semyon Gregoriev, who led the expedition to recover its remains, is quoted as having said? Dan Fisher was one of the first to examine the specimen, and he confirms that parts of it are indeed impeccably well preserved. As to whether it is sufficiently well preserved to contain intact nuclei, we will have to wait and see. I remain skeptical.

AND SO THE SEARCH CONTINUES

It so happened that the two men who appeared suddenly outside of our enclosure on the third day of our ill-fated Taimyr expedition were related to the Jarkovs—the family that found and alerted Bernard of the Jarkov mammoth in 1997. They were Dolgans, a group of people who are indigenous to that part of the Taimyr. While the rest of us were trying to pretend that the sudden appearance of strangers with guns had not nearly caused us to have simultaneous heart attacks, Bernard was inviting them into the enclosure and exchanging hearty handshakes and *bises*. Bernard, it appears, knows everyone in Siberia.

Dolgans are nomadic reindeer herders. During the summer months, they move around the tundra, allowing their large herds to graze. They settle in one spot for a few weeks, until the reindeer have eaten everything in sight, and then pack up and move to the next place. In doing so, they have a chance to scope out pretty much the entire region. If bones, tusks, or mummified mammoths had been exposed when the ground thawed that spring, the Dolgans would know about it. The two men who joined us had seen our helicopter fly in a few days earlier and were curious to know what was going on. So, while the rest of their families were packing up to move on to their next location, the pair set out in search of us.

As the initial shock of the men's surprise appearance wore off, the heaviness that had settled in among the members of our expedition team began to lift and be replaced by the familiar, excited anticipation of what was to come. We gave them all the fish and rice they could eat and apologized for the lack of vodka. When the French couple opened the cooler and pulled out two giant cheeses—a gouda the size of a human head and what must have been three kilograms of brie—the entire crew erupted into laughter. *Of course* a French family working alone in Siberia would have a cooler filled with cheese. Even Pasha, who had managed to inch his face into the enclosure in a desperate attempt to keep the mosquitoes out of his nose, sniffed and flopped his tail onto the tundra. The whole scene was completely absurd, and we were only on day three.

We invited the Dolgan men to stay in our camp for the night and, the next morning, took them back to their families in our outboard-powered inflatable boats. The entertained us for a while; we chatted about the weather, shared some French cheeses, and ate some of their prepared dried fish. We asked whether any of the Dolgans knew of sites that were actively producing bones. They had a few ideas but no strong leads. Then they finished packing up, hooked up their houses and gear to the reindeer, and set off for their next stop on the tundra.

During the rest of the summer, we found only a few scraps of mammoth bone, as well as intact but poorly preserved bones from horses, steppe bison, and woolly rhinos. We later learned that the area we were searching had been covered by ice for most of the Pleistocene, which explains our lack of success. Luckily, before we left Siberia, Ian and I were able to take samples from some extremely well preserved bones that had been collected during previous years' expeditions and that were stored in Bernard's collection in Khatanga, so the trip was not entirely wasted.

These bones did not contain cells with intact genomes. Fortunately, however, perfectly preserved genomes are not critical to de-extinction.

CHAPTER 5

BREED THEM BACK

So, mammoth cloning is not going to happen. No intact genomes will have survived the 3,700 years since the last mammoth walked on Wrangel Island. No mammoth chromosomes will be found that are sufficiently repairable to transform the cells in which they are found into pluripotent stem cells. From my perspective, it doesn't matter how many trips are made to deepest Siberia or how many tunnels are blasted into the permafrost. It's just not going to happen.

Should we just give up? Walk away dejectedly with our tails between our legs? Go back to the rest tent and cry into our mosquito-laden rice soup? Of course not! As it turns out, there are perfectly reasonable, perfectly feasible ways of bringing back a mammoth. Well, of bringing back *kind of* a mammoth. But let us not drown ourselves in the semantic argument just yet. First, the science.

There are two ways to bring an extinct species back to life that are feasible in the present day. One of these is so straightforward that most people probably have not thought of it in the context of de-extinction. The other is more magical, and by "magical," I mean the most-incredible-scientific-advance-in-a-long-while kind of magical. Let's begin with the more straightforward approach.

It is possible to bring an extinct species back right now using technology that our species began to refine some twenty or thirty thousand years ago. It is around this time that we find the first genetic and archaeological evidence of domestication—changing the course of evolution to suit our needs and desires. The approach is not overly sophisticated and requires only a reasonable grasp of basic evolutionary biology. Mainly, the idea is to take advantage of three facts. First, the physical and behavioral characteristics that define an individual—that individual's phenotype—are determined by the sequence of the individual's genome—its genotype—and the interaction of that genotype with the environment. Second, genotypes are passed down from parents to offspring. Third, natural selection can change the relative frequencies of different phenotypes within a population. In the wild, phenotypes that are better adapted to the environment in which the organism lives will become more common than phenotypes that are less well adapted to the same environment.

To bring a mammoth back, we can simply take advantage of nature's own process of genetic engineering. All we have to do is find the hairiest, most cold-tolerant elephants that exist and breed them with each other. After a few generations, we will have created, without calling upon any DNA-sequencing technology at all, an elephant that can live in Siberia.

BACK-BREEDING

Henri Kerkdijk-Otten is a friend of mine who lives in the Netherlands and loves cows. Specifically, he loves large brutish cows that may or may not taste very good and probably don't enjoy being milked. Henri loves aurochs. Unfortunately for Henri, aurochs have been extinct since the middle of the seventeenth century.

Henri, however, has a plan. He will bring his beloved aurochs back from extinction not by finding well-preserved fossils in European forests and not by nuclear transfer but by the comparatively simpler process of selective breeding. His hope is that he can create an auroch by carefully selecting and breeding animals that have physical and behavioral traits reminiscent of the ancient aurochs. After this process of choosing which cow gets to mate with which bull continues for many generations, the aurochs (or at least a close rendition of the aurochs) will be back. They will be able to roam free in the Dutch grasslands, where they will presumably thrive on the ubiquitous tulips.

Aurochs are the wild ancestors of domestic cattle. Around 10,000 years ago, human populations in the Near East and South Asia began farming and taming wild aurochs. Eventually, this gave rise to the two main variations of domestic cattle—humpless taurine cattle and humped zebu. Today, taurine cattle are widely distributed across the globe and belong to familiar-sounding breeds like Holstein, Angus, and Hereford. Zebu tend to be farmed in the tropics, thanks to adaptations that allow them to survive better than taurine cattle in very warm climates. Because domestic cattle are descended from aurochs, much of the genetic diversity that was present in wild aurochs is probably still present in living cattle. It may, however, be distributed among the various breeds. To reengineer an auroch, one simply has to concentrate into a single new lineage all of the auroch-like traits that are present in living zebu and taurine cattle. The end product will not contain the genome sequence of a purebred auroch. It will, however, *look like* an auroch.

The first genetic-engineering experiments performed by humans involved genetic manipulation of wolves, probably gray wolves that lived in Europe as long as 30,000 years ago. It is at this time that we find the first probable evidence of domestic dogs: bones found in archaeological sites that look similar to but distinct from the bones of gray wolves. These early stages of dog domestication were, of course, not hardcore genetic-engineering experiments. Instead, wolves that were more tolerant of humans and humans who were more tolerant of wolves both benefited from a closer association. Just like my own dogs, these first dogs benefited from access to table scraps. The people living in proximity to these early dogs benefited from early warnings of approaching danger, much the same way I benefit from knowing that the mail has arrived. Once the symbiosis was established, humans put genetic engineering to work. Today we have big dogs, small dogs, strong dogs, fluffy dogs, dogs with short legs, dogs with long ears, hunting dogs, herding dogs, dogs that can find people buried in avalanches, dogs that provide life support for people with disabilities, and dogs that can be carried in leopard-spotted purses on trips to the grocery store.

Henri and his colleagues plan to reverse-engineer the domestication process in cattle. Instead of breeding for traits that we tend to associate with domestic animals—tameness and manageability, for example—they want to re-create the wild ancestor of the domestic cow. Beginning with the more "primitive" breeds—including Maremmana, Moronesa, and two Dutch breeds, Limia and Sayaguesa—they have developed a selective breeding program designed to capture the physical and behavioral characteristics of aurochs and, in doing so, create a new cattle breed. The process is known as *back-breeding*, which is a name that highlights the goal: to breed back traits that used to exist and hopefully still exist somewhere in the gene pool of living individuals.

Today's effort is not the first attempt to back-breed the auroch. In the 1920s and '30s, the German brothers Heinz and Lutz Heck, who happened to be directors of the Hellabrunn Zoological Gardens in Munich and the Berlin Zoological Gardens, respectively, were instructed to re-create the auroch. This directive is said to have come from Hermann Göring, who, as an avid hunter, wished to re-create the folkloric prey of Roman hunters. (Although it is unappealing to attribute the first back-breeding experiments to the Nazis, one cannot ignore the timing of this work in interpreting its motivation.) The Heck brothers had the same goal but performed their experiments separately. They each selected different cattle breeds and used these breeds in different crosses. At the time, no scientific reconstructions of the auroch were available, and so neither brother had a particularly good sense of what an auroch actually looked like.

In 1932, Heinz Heck declared his back-breeding experiment a success. A bull was born that he felt looked similar enough to what he believed an auroch should look like that his new bull could be called an auroch. According to Heinz's records (which he stopped keeping after the birth of this bull), the bull was 75 percent Corsican cattle, 17.5 percent Gray cattle, and the remaining 17.5 percent was a mix between Scottish Highland, Podolic Gray, Angeln, and Black Pied Lowland cattle. Selective breeding continued after the birth of this bull, eventually giving rise to what is today known as Heck cattle. Around two thousand Heck cattle are alive today, living in zoos and roaming pastures, mostly in Europe.

Are Heck cattle aurochs? Heck cattle certainly look primitive, particularly to someone who (like the Heck brothers) might not have access to accurate reconstructions of real wild aurochs. Heck cattle have dark coats and long, curved horns, which are two characteristics that were definitely found in wild aurochs. Heck cattle are also more cold tolerant than many other domestic breeds and can survive under relatively poor forage conditions, much as their wild ancestors must have done during Pleistocene glacial cycles. That, however, is about where the similarity ends. Heck cattle are large for domestic cattle, but not as large as the average auroch bull would have been. A Heck bull stands around 1.4 meters high at the shoulder and weigh up to 600 kilograms. An auroch bull, with a shoulder height around 2 meters, would have been taller than the average European man. Also, while the coat color of Heck bulls is similar to what we believe was characteristic of auroch bulls, Heck cows are lighter and more variable in color than auroch cows were. The overall body shape of Heck cattle is also different from that of aurochs, mainly in that it is smaller and, like all domestic taurine cattle, lacks the prominent neck musculature of the wild ancestors. Finally, while the horns of Heck cattle are long, relative to those of domestic cattle breeds, their shape and curvature are somewhat different from an auroch's: they curve slightly too close to the head and point a little bit too far outward.

It is safe to conclude that the Heck brothers did not quite hit the mark. This failure, however, does not spell doom for the present back-breeding project. Today, we know much more than the

Heck brothers knew in the early twentieth century about what traits defined aurochs. We have better descriptions of the various breed phenotypes and a better understanding of the temperaments of these breeds. We have abundant genetic data that help to determine which breeds are the most primitive. We even have ancient DNA data from actual aurochs. Using all of these data, there is little doubt that we will make different and more scientifically justified choices about what animals to use in the backbreeding project, which will eventually lead to the birth of animals that better resemble wild aurochs.

Of course, these animals will not actually *be* aurochs. Not exactly, anyway. Selective breeding is a process by which individuals that display the desired phenotype are bred together to try to replicate that phenotype in the next generation. The phenotype, however, is a consequence of the interaction between genotype and environment. Genetically, the gradual concentration of genes that code for auroch-like traits has to happen by chance. When the gametes—the sperm or egg cells that will go on to become the next generation—are formed, each one contains a shuffled version of that organism's parents' genomes. This shuffling of genetic material, called *recombination*, is an important source of genetic variation within populations. Recombination shuffles genes or parts of genes from mom's chromosome onto dad's chromosome and vice versa. When the sperm or eggs are formed, they will contain some DNA from mom and some DNA from dad. If a phenotype that we want to select is coded for by a gene that came from mom, but the egg that is fertilized contains dad's version of that gene, then the offspring, despite our best intentions, will not display that phenotype.

We can guide the process of concentrating specific traits into a single lineage by selective breeding, but we cannot selectively choose which gametes go on to become that next generation. Some offspring will get the right genes and display the desired phenotype, and others will not. This does not mean that the process will never work. It will, however, be slow. Selecting for multiple traits simultaneously will be particularly challenging, as the genes for each trait need to wind up, by chance, in the same fertilized egg. Despite this, selective breeding is and has been a powerful tool in our species' history, as attested by the variety of forms of domestic plants and animals that we encounter every day. There is no reason why, with sufficient time, resources, and patience, we cannot recover at least some traits of the wild auroch using the selective breeding approach.

As the auroch back-breeding experiment proceeds, I anticipate that the animals will gradually become more and more auroch-like in their physique and behavior. Some auroch traits may, however, never be recoverable from living cattle breeds. The DNA sequence that coded for a particular trait may have been lost, for example, or the trait may have been the product of a genetic interaction with an environment that no longer exists. Some people (myself included) would argue that this does not matter—that by filling the niche of the auroch, even partially, the experiment is a success. De-extinction purists will, however, never be satisfied with a back-breeding product, because the result will always be something *new*, not something old. Auroch 2.0 will not be an auroch. Not precisely, anyway.

IS SIMPLER NECESSARILY BETTER?

One advantage to back-breeding as a means of de-extinction is that it relies so little on molecular biology technologies. Genomes don't have to be sequenced, genes don't have to be identified, and genetic variants don't have to be linked to specific traits. The gradual transition from one form to another happens without embryonic stem cells and long hours spent in a lab. And the results of the experiments are validated qualitatively: does it or does it not look more like an auroch?

The simplicity of back-breeding, however, may also be its downfall. While traits such as dark coat color, long forward-facing horns, or strongly expressed neck and shoulder musculature may emerge in the population after some generations of selective breeding, the genes that code for these traits once the traits reemerge may be different genes from those that coded for the same traits in the extinct species.

Does it actually matter? If we want long forward-facing horns, and the bull has long forward-facing horns, does it really matter what *specific* genes are making it happen? It might matter. Genes don't always, or even often, have just one function. A gene that makes curved horns might have other consequences on the resulting cattle phenotype that we don't want. It might make their skull slightly differently shaped, for example, or somehow influence the shape or texture of their hooves. In addition, genes don't act in isolation but instead act in concert with other genes that are also expressed in the cell.

An example of an interaction between genes that is used in introductory biology classes is the way that coat color in horses is determined. Horses have a single gene that determines whether their coat will be red or black. The dominant allele produces a black coat and the recessive allele produces a red coat. If this gene acted alone, individuals that carried either two copies of the dominant allele or one dominant and one recessive allele would have black coats and individuals that carried two copies of the recessive would have red coats. However, there are many different types of red or reddish horses. This comes about because of yet another gene—the cream dilution gene—that modifies the expression of the red alleles. A horse that has two copies of the recessive red allele can be chestnut colored, palomino, or even white or cream colored, depending on how many copies of the cream dilution allele it carries.

While not all interactions among genes are known, and very few are well understood, this does not mean that selective breeding for specific traits is impossible. Through multiple generations of back-breeding, using different crosses involving different individuals or different breeds, the right combination of genes, or at least combinations of genes that provide the right phenotypes, may eventually be discovered. How long it will take depends on several factors, including how many traits are being selected, how easy the animals are to breed, and how long it takes to go from one generation to the next.

TOO SLOW FOR SUCCESS?

The generation time of cattle is short compared to many species. Female cattle can breed for the first time when they are between one and two years old, and gestation takes around nine months. A selectively bred individual can be born, develop into an adult, get pregnant, and give birth to the next generation all within two to three years. While not a breakneck pace, one can imagine how a selective breeding program could progress reasonably quickly in cattle.

Progress would be much, much slower for some of the other candidate species for de-extinction. For example, male elephants begin making sperm between ten and fifteen years old, and female elephants in the wild will become pregnant for the first time around age twelve. Gestation time in elephants is between twenty and twenty-two months. That means there would be a fourteen-year wait between when the first selectively bred offspring is born and when that offspring can produce the next generation. At that pace, only five generations could be produced in a human lifetime. There must be a better way.

Of course there is. An easy way to minimize the time it takes to selectively breed a trait into a lineage is to make sure that every individual in the next generation contains the target trait. This is not possible with back-breeding, where the offspring of two parents may or may not inherit the target trait or traits. However, new technologies—specifically, the genome engineering technologies that are behind the second presently feasible (and the more magical) pathway to de-extinction—make it possible to edit the genome directly. By manipulating the DNA sequence in a cell and then using that cell to create living individuals, we can be certain that the target trait is present in the next generation. We can make the entire process of resurrecting extinct traits in living species move along much more quickly and efficiently.

For example, we know that mammoth hemoglobin—the protein in red blood cells that takes up oxygen in the lungs and then distributes it via the circulatory system to the rest of the body—differs from elephant hemoglobin by exactly four mutations.

These four differences modify the performance of the hemoglobin by making the mammoth version more efficient than the elephant version at delivering oxygen when the temperature in the body is very low (think mammoth feet standing in the snow).

We will not find a living elephant that has the mammoth version of these hemoglobin genes. The common ancestor of mammoths and living elephants lived in the tropics, and adaptations to life in the cold would have evolved in mammoths only after the mammoth lineage diverged from the Asian elephant lineage. Since all mammoths are extinct, there are precisely no individuals alive who have these particular genes. In order to create an elephant that makes mammoth hemoglobin, we will have to make the mammoth version of those genes from scratch and then somehow insert that version of the gene into an elephant cell. We can do that.

CHAPTER 6

RECONSTRUCT THE GENOME

In 2010, J. Craig Venter made life from scratch. He and his team synthesized the complete genome of a tiny, free-living bacterium, which they named *Mycoplasma mycoides* JCVI-syn1.0, and transplanted it into a recipient cell whose own genome had been removed. In addition to stringing together all of the genetic bits and pieces required to make the genome function (which it did) and the cell replicate (which it did) they added watermark sequences—the names of the researchers involved in the project translated into a genetic code—to distinguish the synthetic genome from the real genome on which the copy was based.

Venter and his team began the life-creating process by learning the complete genome sequence of a living *Mycoplasma mycoides* bacterium. This digitized genome, which was nothing more than lines of text stored in a file on a computer's hard drive, became their blueprint for constructing life. They chose this particular bacterial genome because it was short—a little over a million base-pairs long—and because the bacteria grew quickly, which meant that it would not take a long time to complete the experiment.

One million base-pairs is a very small genome, even for a bacterium. It is not, however, sufficiently small to synthesize all at once. When strands of DNA are produced in a lab, machines do this by stringing together single nucleotide bases—the As, Cs,

Gs, and Ts that make up entire genomes—in order. The longer the fragment is, the more mistakes will be made during synthesis. If this bacterium was going to be able to survive and reproduce, they would have to make a synthetic genome that was as identical to the blueprint genome as was possible.

To get around the problem of synthesizing long fragments, Venter's team designed a four-step process to construct the complete genome. First, they synthesized, one base-pair at a time, 1,078 fragments of DNA that were each 1,080 base-pairs long. These fragments were sufficiently short to construct reliably in the lab, but also long enough to each contain unique identifying information that would be used to orient them correctly in the final genome. Then, taking ten fragments at a time that they knew were adjacent to each other in the blueprint genome, they inserted these smaller fragments into yeast cells and allowed the yeast's cellular machinery to stitch these fragments together. That process provided 100 fragments of bacterial DNA that were each around 10,000 base-pairs long. They then took ten of these at a time and stitched them together, making eleven fragments of around 100,000 base-pairs long. Finally, they stitched these eleven fragments together to create a single million-base-pair-long bacterial genome. They removed this genome from the yeast cell and inserted it into a bacterial cell, where it began to make all the proteins necessary for life. The entire process took fifteen years and cost more than US$40 million.

Creating the first synthetic life was an awesome accomplishment. It does not, however, get us any closer to being able to create mammoths or passenger pigeons. First, bacteria are prokaryotes, which means they lack a nucleus. Because of this, Venter and his team were able to skip an important and as yet unsolved step in the life-creation process: they did not have to assemble a genome comprising multiple distinct chromosomes *within* a nuclear membrane, as would have been necessary to create a eukaryote. Until somebody figures this out—and I'm watching this space with a close eye on the J. Craig Venter Institute—there will be no mammoths or passenger pigeons running around with completely synthesized genomes. Second, bacterial genomes are small. The mammoth's genome is more than four billion base-pairs long. Birds tend to have much smaller genomes than mammals, but even their genomes are mostly more than one billion base-pairs long. Not all of these base-pairs code for genes that make proteins, but we still don't really know how much of the other stuff in a genome is essential for life. More importantly, we do not and probably cannot know the complete sequence of any extinct species' genome. Even if scientists were to discover a means to synthesize an entire eukaryotic genome within the nucleus of a cell, we may never have the template for that synthetic genome.

Let's take a closer look at the mammoth. Over the years, ancient DNA scientists have sequenced billions of base-pairs of mammoth DNA from dozens of mammoth bones and other remains. The fragments of DNA that are recovered from these tend to be short—somewhere in the range of thirty to ninety base-pairs long—and they are damaged, as is expected for very old DNA. Returning to the puzzle analogy from chapter 2, our mammoth DNA fragments are the puzzle pieces and the picture on the box top is of the African elephant; while we know from comparing their mitochondrial DNA sequences that the Asian elephant is more closely related to the mammoth than the African elephant is (figure 9), thus far only the African elephant nuclear genome has been reconstructed, so only the African elephant genome is available to act as a guide. Also, only about 80 percent of the African elephant genome is known, so there is something not quite right about the photo on the box top. In essence, we have billions of microscopic, slightly misshapen puzzle pieces and a slightly blurry photographic key that solves a different puzzle.

The easiest pieces of the puzzle to put together will be those that come from the most conserved regions of the genome. These are the parts of the genome where mammoths and both species of living elephant, and indeed all mammals, are identical or nearly identical. We should also be able to assemble pieces that come from genomic regions in which the mammoth and the African elephant are similar but not perfectly matched. The hardest pieces to assemble will be those that come from regions of the

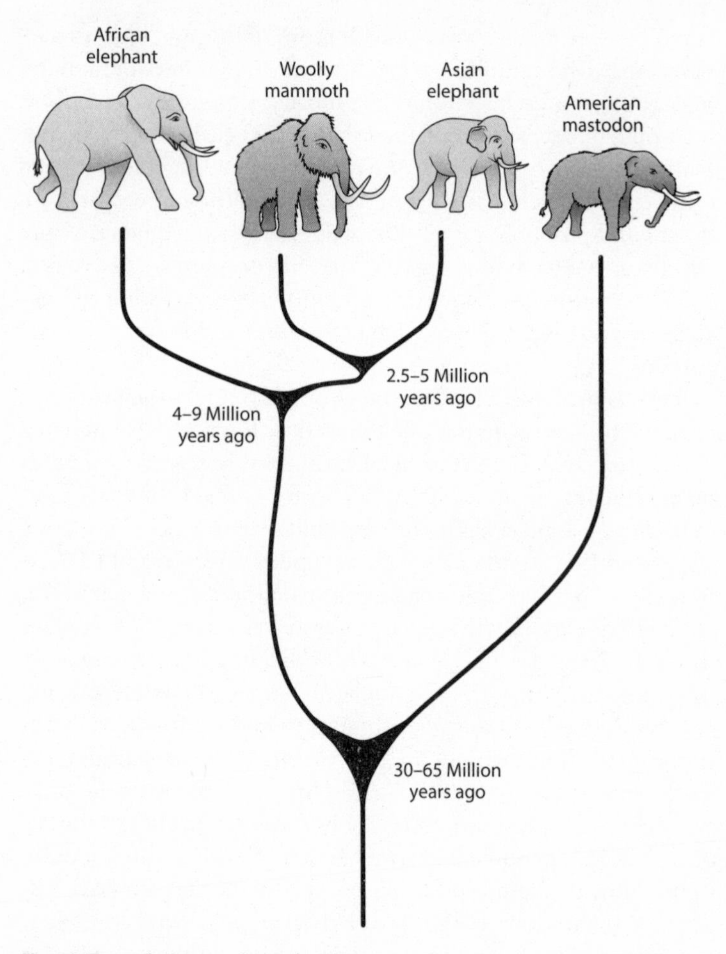

Figure 9. The evolutionary relationships between mammoths, mastodons, and Asian and African elephants, based on the fossil record and their mitochondrial genome sequences.

genome in which the mammoth and the African elephant are very different. These differences might be due to reshuffling of genes or even duplication or deletion of genes.

In the book *Jurassic Park*, scientists filled in the bits of the dinosaur genome that they were unable to sequence with frog DNA.

Similarly, we might solve this problem by simply filling in the holes of the mammoth genome with elephant DNA. This does not seem to me to be a very good idea, however. The common ancestor of mammoths, Asian elephants, and African elephants lived around four million years ago. This means the mammoth is separated from both the Asian elephant genome and the African elephant genome by more than eight million years of evolutionary change—a long time in which to collect evolutionary differences. Some of the hardest parts of the genome to assemble will be those regions that have changed in mammoths since their divergence from the other elephants. Arguably, these will be among the most important genomic regions to change in order to create an elephant that looks and acts like a mammoth instead of like an elephant. For the purposes of de-extinction, these might be the most critical regions of the genome to get right.

When assembling genomes from living species, the best way to assemble these trickiest regions is to sequence very long fragments of DNA. By long fragments, I mean fragments that are thousands to hundreds of thousands of base-pairs long. This is hard to do, and enormous biotech budgets are spent each year trying to solve this problem. Unfortunately, ancient DNA does not survive in long strands—our fragments mostly contain fewer than one hundred base-pairs and frequently very many fewer than that. So, even if breakthroughs in sequencing technologies emerge over the next few years that allow very long fragments to be sequenced, these will be of little use in sequencing the genomes of extinct species.

The good news is that the cost of sequencing DNA continues to decrease, which means that we can generate more and more sequences from each ancient sample without completely breaking the bank. Also, our ability to recover DNA from fossils is improving. While these sequence fragments will always be short, the quantity of recoverable fragments will increase. We might also get lucky and find ancient samples—preserved in frozen arctic soils, for example—that do retain fragments of DNA that are many hundreds of base-pairs long, although we are extremely unlikely ever to find remains that retain fragments that are thousands or tens of thousands of base-pairs long. Finally, computational approaches to piecing together fragments of DNA without a closely related guide genome are also improving, which allows better assemblies of ancient genomes from increasingly divergent species.

The truth, however, is that zero mammalian genomes are sequenced entirely to completion. This includes the human genome, although the ecstatic claims of having done so more than a decade ago would certainly suggest otherwise. The truth is that there are some regions of the human genome that remain unsequenced to this day and cannot be sequenced using any existing sequencing technology.

Genomes comprise two components: the euchromatin, which is the component in which the genes are found, and a highly repetitive and tightly condensed component called heterochromatin. In the human genome sequence, there are still some (very) small unsequenced gaps remaining in the euchromatic portion of the genome, but these amount to less than 1 percent of the human genome. The other, larger missing component is in the heterochromatic sequence. Heterochromatin makes up about 20 percent of the human genome and, thanks to its highly repetitive nature, is the most difficult portion of the human genome—or any genome—to sequence. Heterochromatin probably plays important roles in regulating gene expression, in directing the segregation of chromosomes during cell division, and in determining where the different chromosomes live within the nucleus. Because it is so difficult to sequence using existing technologies, however, we know very little about it compared with what we know about the euchromatic portion of genomes.

Heterochromatin will be no simpler to sequence from an ancient sample than it is from a living human. In fact, sequencing heterochromatin from degraded samples is likely to be extra challenging compared with sequencing from samples of living organisms, thanks to the fragmented nature of ancient DNA. It remains to be known whether this is an important roadblock to de-extinction.

Because we cannot know the complete genome sequence of an extinct species, synthesizing a complete genome from scratch would not be an option for de-extinction even if it were to become possible to re-create synthetic eukaryotic life. I strongly believe, however, that synthetic biology *is* the way to bring extinct species and traits back to life. While we cannot synthesize an entire genome, we can synthesize fragments of DNA. What if we could use these fragments of DNA to engineer extinct species back to life?

CUT AND PASTE A MAMMOTH

George Church is a professor of genetics at Harvard Medical School and is the leading partner in another mammoth de-extinction project, one that is markedly different from those that rely on finding intact cells in the Siberian permafrost. George is using genome engineering to resurrect a mammoth, which is, as I said, one of the two presently feasible methods for resurrecting extinct traits.

I first met George at the Wyss Institute in Cambridge, Massachusetts, in 2012. He was hosting a mini-conference that was organized by Ryan Phelan and Stewart Brand of the Long Now Foundation as part of their new nonprofit undertaking, Revive & Restore. The conference was notionally about a project to bring back the passenger pigeon, and as the scientist with the largest collection of passenger pigeon bits in her ancient DNA lab, I was invited to attend. Also attending were conservation biologists, including Noel Snyder from the US Fish and Wildlife Service, who has spent many years of his life working on the project to save the California condor, and scholars of bioethics, like Hank Greely, a Stanford law professor who specializes in the social and ethical implications of biotechnology. The conversation was intense and at times angry, but it was tremendously useful: it was at this mini-conference that I realized how de-extinction was going to happen.

George Church is one of my favorite scientists. There are few people in the world who successfully straddle the gulf that separates genius and madness, and he is one of them—probably because his genius far outweighs his madness. George Church is one of the most inventive minds in genomics, a fact that is most apparent in the excessively long lists of biotech partnerships that appear at the ends of his papers and presentations.

At the meeting in 2012, George presented his plan for bringing a mammoth back to life as a model for what could and should be done for the passenger pigeon. His plan involved using new (and awesome) technology to change the elephant genome, bit by bit, into a mammoth genome. His plan can most simply be summarized as a cut-and-paste job. I'll describe it in much more technical detail later, but for now, here are the basics.

First, we collect a few (or many) well-preserved mammoth remains, extract DNA, and assemble a genome. We then compare that genome to the genome sequence of an Asian elephant, and identify the parts of the genome where the mammoth sequence is different from the elephant sequence in some important way. This provides us with our plan: we will edit the elephant genome so that it looks like the mammoth genome in those specific places.

Second, we synthesize strands of mammoth DNA that match the genomic regions that we want to change. We do this by stringing together As, Cs, Gs, and Ts and following the template provided by our assembled partial mammoth genome. This provides us with strands of DNA that we will later paste into the elephant genome. These synthesized fragments could be very short (only a few base-pairs long) or somewhat longer (several hundred or possibly several thousand base-pairs long), but they are much shorter than the length of a chromosome and are certainly within reach of what is feasible in the present day.

Third, we engineer a tool—let's call it "molecular scissors"—whose job it will be to find and bind to precisely the sequence within the elephant genome that we want to change. There are several such tools, all of which I will describe later.

Fourth, we deliver the synthesized strands of mammoth DNA and the molecular scissors into the nucleus of an elephant cell. The molecular scissors locate the precise spot in the elephant genome where the edit is to be made, bind to it, and cut the strand of DNA in half. Because having broken DNA is bad for the cell, cellular machinery has evolved that will fix exactly this type of DNA damage. This cellular machinery kicks into action and fixes the broken strand by pasting the mammoth version of the sequence in place of the elephant version.

Fifth, we measure the success of the cut-and-paste by designing an experiment that allows us to learn whether the cells are now expressing the mammoth gene and not the elephant gene. This step allows us to identify those cells that have been edited and then to measure how, if at all, the edits change the phenotype of the cell.

Finally, those cells in which all the cut-and-paste jobs have been successful are used in nuclear transfer to create living organisms with selectively engineered genomes.

I think I can safely speak for the others attending the meeting in saying that we were, as a group, rather taken aback by how real and achievable George's presentation made de-extinction feel. His approach seemed simple, even elegant. Could it be true that living, breathing mammoths really were within reach within the time frame proposed by Professor Iritani (although not by the same means)?

At the time, George had not yet started manipulating elephant DNA. The mammoth genome was still in the very early stages of assembly, and, as such, it was not entirely clear what parts of the elephant genome should be targeted for editing. We were also still in the process of sequencing the passenger pigeon and its closest living relative, the band-tailed pigeon, so we, too, had little idea about what we might actually change in the band-tailed pigeon to make a passenger pigeon. This presentation, however, clarified what our goal should be. And, more importantly, this goal appeared achievable. We did not need to sequence the complete genome. We simply had to figure out, somehow, which parts of the genome were important and sequence those.

MOLECULAR SCISSORS AND ENZYMATIC GLUE

While genome editing as described by George Church sounds pretty straightforward, the process—unsurprisingly—faces significant technical challenges. To be successful, genome editing has to be *specific*. Nobody wants molecular scissors to go around wantonly chopping up a genome and randomly inserting DNA. This would not only *not* have the desired effect on the phenotype of the cell (or the resulting animal), but nonspecific chopping up of DNA is actually toxic to the cell. It causes genomic instability and often cancer.

Key to the success of genome editing has been the discovery and development of different types of *programmable* molecular scissors. Programmability allows specificity, which means we can make the cuts we want to make where we want to make them, and we can avoid making cuts that kill the cell.

For the last decade or so, two types of programmable molecular scissors dominated the field (figure 10): *zinc finger nucleases* (ZFNs), and *transcription activator-like effector nucleases* (TALENs). ZFNs and TALENs are similar in that they are both hybrid molecules made up of two distinct parts. The first part is a protein that recognizes and binds to the part of the genome that is to be edited, which is sometimes called the "arm." This is the programmable part—each zinc finger recognizes a specific sequence of three nucleotides, and each transcription activator-like effector (TALE) recognizes a single nucleotide. Chains of zinc fingers or TALEs are strung together synthetically so that each chain recognizes a specific sequence of DNA. The second component of the hybrid molecule is the nuclease. The nuclease is the scissors that actually make the cut. The nuclease is attached to one end of the chain of zinc fingers or TALEs. Two hybrid molecules are synthesized for each edit that is to be made: one that finds and binds to the DNA sequence that lies upstream of the target site

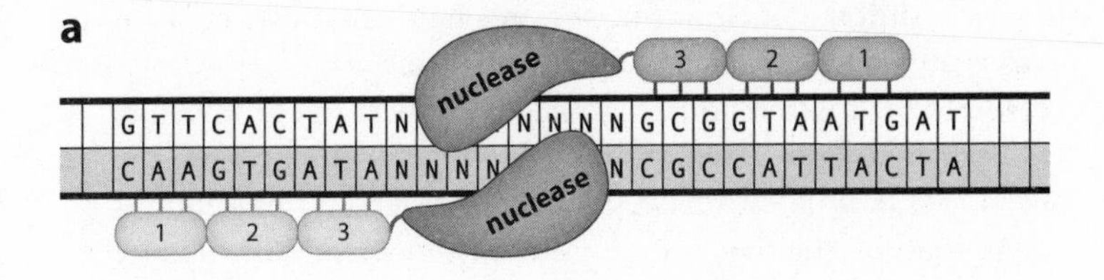

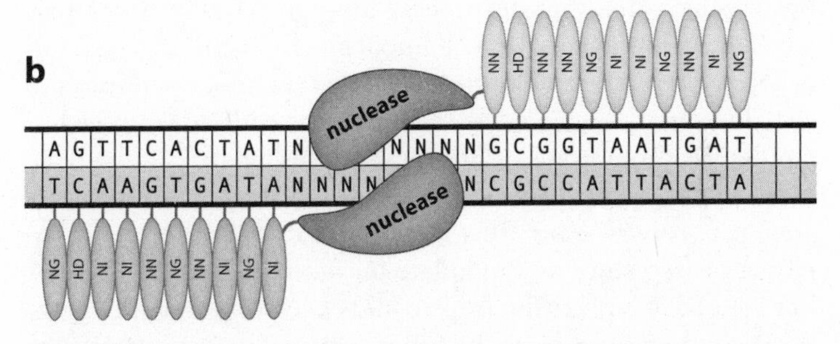

Figure 10. Zinc finger nucleases (ZFNs) and transcription activator-like effector nucleases (TALENs). Each finger in a ZFN recognizes a specific sequence of three nucleotides, whereas each transcription activator-like effector (TALE) recognizes a single nucleotide. The arms are created by linking a nuclease and specific sequence-recognizing fingers or TALEs together so that their sequence matches the genomic sequence to which they are intended to bind.

and another that binds to the DNA downstream of the target site. When both molecules have located exactly the right spot in the genome and have bound to it, the nuclease makes a cut.

Making the correct cut is only the first half of the cut-and-paste process. The second half of the cut-and-paste process involves tricking the cell to replace the elephant version of the sequence with the mammoth version of the sequence as it repairs the newly broken strand of DNA.

Normally, cutting both strands of DNA would be lethal to the cell. If only one strand were cut, the cell's repair machinery could fill in whatever sequence was lost using the other strand as a template. If both strands are cut, it is less obvious how the cell would know how to replace any missing sequence.

Two different cellular-repair mechanisms have evolved to solve this problem. The first is called *homologous recombination*. Because there are two homologous copies of each chromosome in the cell (one from mom and one from dad), one of these can be used as a template to fix errors in the other. In homologous recombination, the two homologous chromosomes line up next to each other and recombine, allowing the cell's repair machinery to use the intact chromosome's sequence as a template to fix the break. The cut-and-paste process aims to harness this repair mechanism but to trick the cell into using the synthesized strand of DNA (here, the mammoth DNA that was delivered into the cell along with the molecular scissors), instead of using the homologous chromosome, as the template for repair.

The other mechanism for repairing double-strand breaks is *non-homologous end joining*. This mechanism does not require a homologous sequence as a template for fixing the strand of DNA but instead just glues the broken ends back together. This is *not* the pathway that we want the cell to follow if we want to change the DNA sequence, but it is a pathway that is often used by the cell. Thus, one challenge that remains is to develop a way to control which of these pathways the cell uses to repair the DNA. At the moment, only a fraction of cells that are edited will end up with the new version of the gene pasted in the right place after the break is repaired.

ZFNs and TALENs have already proved to be tremendously powerful molecular tools. ZFNs have been used to fix mutations that are known to cause genetic disease in humans by directly editing the genome sequence in patient-specific stem cells. These modified stem cells can then be transplanted into the patient to cure the disease. ZFNs are even being used to develop a cure for HIV/AIDS by editing the CCR5 gene, which codes for the protein that HIV uses to enter T cells, into a version of CCR5 that the virus cannot use. Genome editing has also been used to insert herbicide-resistance genes into corn and tobacco and to alter cow genomes so that they produce the human version of various blood and milk proteins.

Genome-editing applications of ZFNs and TALENs are limited mainly by the need for specific targeting within the genome, which it turns out is pretty difficult to control. Longer probes made by linking more zinc fingers or TALEs together provide increased sequence specificity, but longer proteins are harder to deliver into the cell. Also, making the probes is a painstakingly difficult process that often requires months or years of trial and error. These are all problems that are encountered when working with organisms with long histories of experimental manipulation in molecular biology labs. If these methods are to be applied to de-extinction, the experiments will include species whose genome sequences are *not* known and that have *never* been used in molecular biology research, compounding the difficulty of making the experiments work. Certainly, these genome-editing tools have potential when it comes to de-extinction. Digging deeply into exactly how they work, however, provides a somber reality check.

A CRISPR VIEW OF DE-EXTINCTION

Around the same time as our Harvard meeting, a new molecular tool appeared in the genome-editing toolbox. This new tool, called CRISPR/Cas9, was first discovered for its role in providing immunity to bacteria by learning a pathogen's DNA sequence and later targeting and destroying that sequence. Harnessing this system for genome editing provides two key advantages over ZFNs and TALENs. First, the programming is much faster—there is no longer a need to link fingers or TALEs together by trial and error. Second, much longer sequences can be used, which provides a tremendous increase in specificity. The relative ease and simplicity with which genome editing can be achieved using this system hints that another revolution in biology—similar to the one that came about when PCR was first developed—may be just around the corner.

Here is how it works. When a pathogen invades a bacterial or archaeal cell, the pathogen genome is recognized and chopped into small pieces. Some of these pieces become incorporated as "spacers" into a molecule called a CRISPR, or *clustered regularly interspaced short palindromic repeats*. In this manner, these bits of pathogen are integrated into the bacterial genome and stored for future use. To defend itself against invading pathogens, the cell transcribes the CRISPR and chops it up at the repeats, releasing the spacers, which, remember, are sequences of pathogen DNA. The transcribed spacers are taken up by Cas9 proteins, which then scan the cell for DNA that matches the sequence of the spacer as a means to find and destroy invading pathogens.

To translate the CRISPR/Cas9 system into one that is useful for genome editing, imagine that, rather than grab bits of pathogen DNA and use these sequences as probes to search for potentially invading pathogens, the Cas9 molecules bind to a sequence that we design and use this to search for the part of the genome we wish to edit (figure 11). This becomes a highly efficient, highly precise way to locate specific parts of the genome. We design and synthesize CRISPR-RNAs, which are analogous to linked-together zinc fingers or TALEs, to find a precise part of the genome. When the CRISPR-RNAs find that part of the genome, Cas9, which is analogous to the molecular scissors in ZFNs and TALENs, makes the cut. After this, the standard DNA repair processes take place and—we hope—our edits are incorporated into the genome sequence.

In addition to gains in both speed and specificity, the CRISPR/Cas9 system also provides an increase in efficiency when we want to make multiple changes at the same time. Cas9 and the synthesized CRISPR-RNAs are not physically linked, which means that many different CRISPR-RNAs can be delivered into the cell at once. Each of these will be captured by Cas9 and used to find (and cut) different parts of the genome.

George Church's team at the Wyss Institute is among the research groups that are leading the development of the CRISPR/Cas9 system for genome engineering. Most people in his lab are thinking about applications of CRISPR in personalized medicine, or about refining the technology so that it is possible both to insert longer fragments of DNA and to perform multiple edits

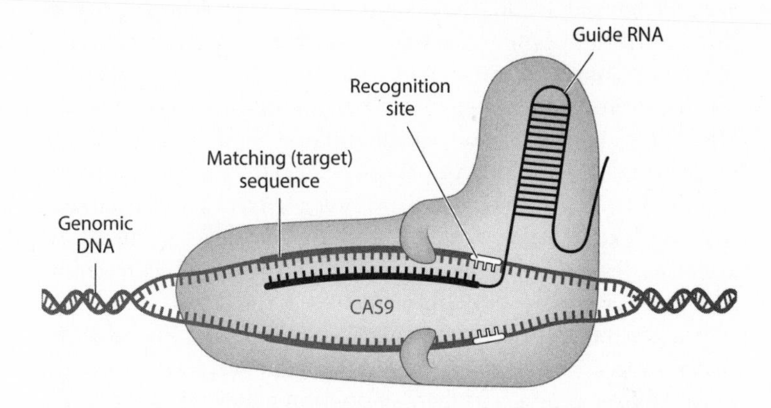

Figure 11. CRISPR/Cas9. Long strands of DNA that match the region of the genome to be edited are synthesized and used to construct CRISPR-RNAs (the dark strands of DNA). These are delivered into the cell with Cas9. Once inside the cell, Cas9 picks up a CRISPR-RNA, which guides the entire complex to the correct place in the genome (the light strands of DNA), and the cut is made.

of different parts of the genome at the same time. But tucked away, in a dark corner of his lab (this is how it is arranged in my imagination, anyway), is a small team of postdocs with a mammoth-sized goal: the self-named *mammoth revivalists*. Every month, labs involved with Revive & Restore connect via teleconference to catch up on progress in our active de-extinction projects. The mammoth revivalists consistently put the rest of us to shame. We are still assembling the passenger pigeon genome and trying to figure out what we might want to change. They've decided not to wait for the mammoth genome to be finished before proceeding at full bore. Beginning with a few mutations that we're aware of—the differences between mammoth and elephant hemoglobin—and a few good guesses, they are cutting and pasting their way to a mammoth.

The mammoth revivalists' de-extinction plans are relatively subdued for now. No Asian elephant cells were available when they got started, so they are editing African elephant genomes in African elephant cells. Also, for now, they are working with a type of skin cells—fibroblasts—and not stem cells, again because that was the only type of cell that was available. They have a separate line of ongoing research to try to make stem cells from the elephant fibroblasts, which has had limited success thus far. Once they have successfully created stem cells, they intend to use these to create different types of cells that can then be used to test whether or not their edits have been successful. No one is talking just yet about actually making these cells into a living mammoth. For now, the goal is to edit the genome and grow cells containing the edited genome in tiny plastic dishes in the lab.

The team hopes to edit the African elephant genome in a way that produces two specific phenotypic changes. First, they will make all four changes in the hemoglobin genes that are known to differ between elephants and mammoths. This should produce cells capable of making mammoth-like hemoglobin. If they can make these changes in hematopoietic stem cells—stem cells that differentiate into different types of blood cells—they will be able to measure directly the oxygen-carrying potential of the resulting red blood cells, which should tell them whether their experiment was a success. They're also hoping to create cells capable of growing what George calls "the thickest, most luxurious, mammoth-like hair." This is a tougher task, though, because no one is certain which (or how many) genes are involved with making thick, luxurious mammoth hair. For the moment, George is content to make an educated guess, based on which genes are believed to be associated with hair phenotypes in other species.

This, of course, is just the beginning. Once it is clear that extinct phenotypes can be genetically engineered into the cells of living species, de-extinction will be off and running. But what precisely will the resulting animal be? How many changes do we have to make in order to call an elephant a mammoth? Is it possible to make every single change that might differ between the two genomes? If not, what *should* we change?

CHAPTER 7

RECONSTRUCT *PART OF* THE GENOME

Here is my prediction: Within the next couple of years, George Church and the mammoth revivalists will succeed in transferring at least one mammoth gene into an elephant stem cell. They will use that stem cell to produce cells that express the newly inserted mammoth gene. They will carefully measure their success by designing a smart experiment to show that the gene is now producing mammoth proteins rather than elephant proteins. When they see positive results, indicating that they have, indeed, engineered a mammoth gene into an elephant cell, they will announce this success with deserved pride. It will be an astonishing achievement.

No elephants will have been harmed in the process. No elephants will have been *involved* in the process, other than to donate blood during a routine veterinary visit. No female elephants will have been subjected to any experimental manipulation whatsoever. No one will have performed nuclear transfer on an elephant. No baby elephant whose genome contains mammoth genes will be gestating anywhere.

The press will not hear any of the above caveats, however. The headlines will read: "The Mammoth Is Back." "Extinction Is No Longer Forever." "Scientists Create Woolly Mammoth in Test Tube." It will be the biggest, most exciting, scariest, most wonderful, most terrible thing to happen in recent memory. There will likely be widespread announcements of dire consequences as well as excitement and some hysteria.

But there is no need, really, to speculate on how people will react. We can simply look to recent history.

THE MAMMONTELEPHASE OF DE-EXTINCTION

On April 23, 1984, the following appeared in the *Chicago Tribune,* tucked neatly away among the inside pages. The headline read: "A shaggy elephant story." The full article is reproduced here, with permission:

> When a species becomes extinct, we expect it to stay that way. Scientists in America and the Soviet Union have upset that seemingly safe assumption by "retro-breeding" a hybrid animal that is half elephant and half woolly mammoth, The story starts in Russia, where Dr. Sverbighooze Yasmilov of the University of Irkutsk was able to extract the nuclei from egg cells taken from a young mammoth that was found frozen in Siberia. *Technology Review* reports that he sent the material to the Massachusetts Institute of Technology, where Dr. James Creak mixed the DNA from the cells with elephant DNA. Woolly mammoths, which roamed Europe until they died out 10,000 years ago, have 56 chromosomes; elephants, their near-relations, have 58. Based on Creak's success, Yasmilov decided to try to fuse the nuclei from the mammoth's egg cells with sperm from an Asian elephant. The experiment produced eight fertilized eggs, which were implanted in Indian elephants. Six miscarried, but two hybrid animals—males that are probably sterile—were born. The hybrids, which some call "mammontelephases," are covered with yellow-brown hair and have jaws that are similar to the mammoths.

The tiny story was picked up and distributed by the *Chicago Tribune*'s news service, and versions of it appeared in more than

350 newspapers within the following days. It even appeared in a nationally circulated Sunday supplement, where it no doubt received the widest potential readership.

Not one of the newspapers that picked up and ran the story bothered to check the facts. If anyone had bothered to contact the author of the report mentioned in the *Technology Review,* for example, or had tried to talk to any of the scientists involved in the research, they would have made a startling discovery: the whole thing was a joke. The scientists did not exist. The project did not exist. The story was meant to be a parody of science, written by a talented undergraduate student to fulfill a science writing assignment. The story was published in the *Technology Review* in celebration of All Fools' Day. The article, which is on page 85 of the April 1984 edition of the *Technology Review,* concludes with the name of the student author—Diana ben-Aaron—and the date—April 1, 1984.

Perhaps those at the *Tribune* and the many other newspapers that decided to run the story were simply too excited about the possibility of mammoth de-extinction to notice the date or to question the authenticity of the report (including the unlikely collaboration between Soviet and American scientists at the peak of the Cold War). Or perhaps they didn't get the joke.

The fictional piece by ben-Aaron was prescient in many ways. She predicted, for example, the poor success rate of nuclear transfer, despite writing the article more than twelve years before of the birth of Dolly at the Roslin Institute. She predicted that the Asian elephant would be used as a surrogate, although it would be more than two decades before we knew with certainty that the Asian elephant is more closely related to the mammoth than the African elephant is. She also anticipated and attempted to defuse some of what the public would fear about de-extinction. For example, she foresaw that containing these new creatures—not allowing them to escape and breed with the wild elephant population—would be a key concern. As Michael Crichton would do six years later, she invented a mechanism by which breeding of the cloned animals would not be possible without human intervention. While Crichton's dinosaurs were all female and therefore unable to reproduce, ben-Aaron's mammontelephases were all infertile males. She made them infertile by giving them an uneven number of chromosomes. With mismatched numbers of chromosomes, they, like mules, would be sterile.[1]

The reaction to the press coverage of ben-Aaron's fictional article was fast, heated, and mixed. Some people celebrated, either because they were amused by the display of obviously poor journalism, or by the parody itself, or because they didn't know it was fake and simply were excited that a mammoth had been brought back from the dead. Others were angry, either because they felt that the parody was improper or unfair or because they didn't know it was fake and were really annoyed that scientists would do such a terrible thing as bring a mammoth back from the dead.

The reaction was, in fact, much like the reaction I anticipate when the mammoth revivalists publish the first evidence that their genome-engineering project has been a success and that edited elephant cells may—at some point in the future—be used to make edited elephants. Of course, the 1984 scenario was entirely fabricated. The hypothetical headlines of our future will reflect actual science going on in an actual cutting-edge research lab at one of the most respected research institutions in the world.

In 1984, those who read and believed the story in the *Chicago Tribune* or elsewhere came away with one message: a mammoth had been brought back to life. That was, however, *not what the article said*.

The headlines to come when the mammoth revivalists produce the first mammoth-flavored elephant cell are likely to be more spectacular than the subdued title of the *Tribune* piece. Careful journalists are unlikely to omit the fact that very little of the elephant genome is actually changed; however, this fact will be conveniently brushed aside to make way for impassioned and melodramatic commentary reflecting on the central message of the piece: a mammoth will have been brought back to life.

Except it *still* won't be true.

Plate 1. Martha, the last known living passenger pigeon, in her enclosure at the Cincinnati Zoo in Ohio, USA. Photo courtesy of the Wisconsin Historical Society, WHI-25764.

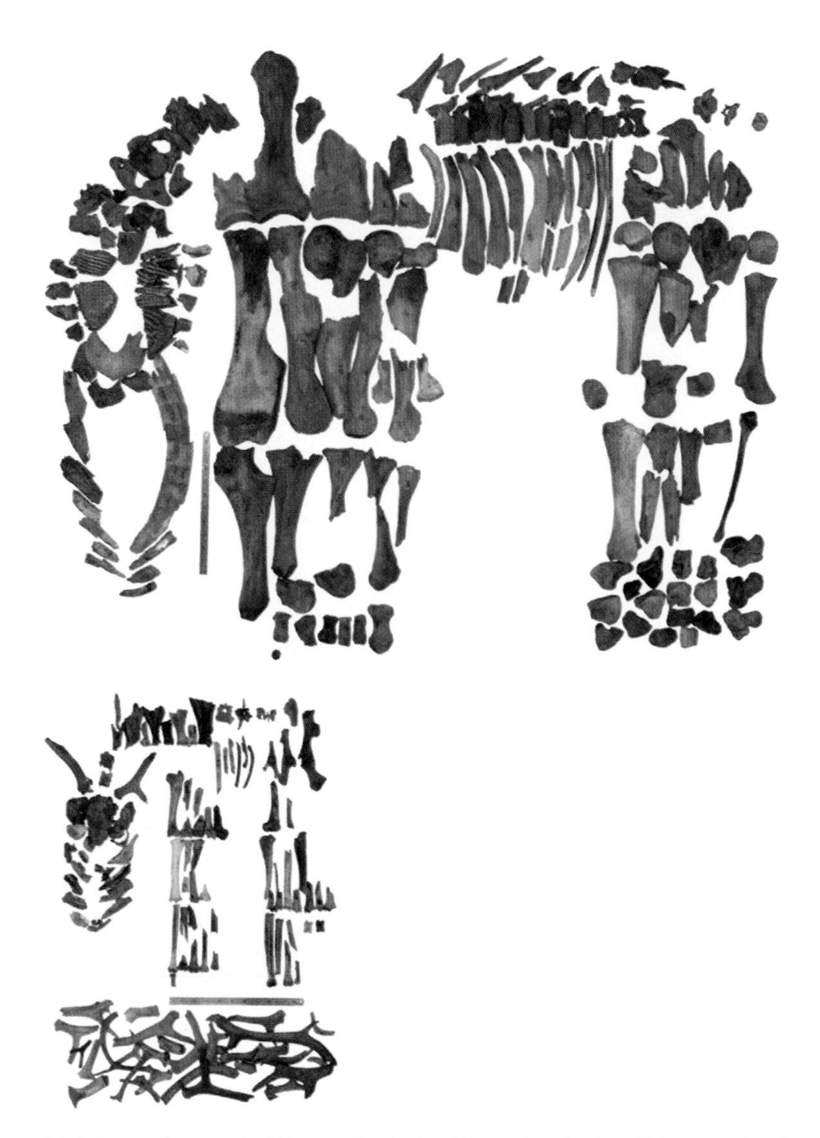

Plate 2. Bones of mammoths (*this page, top*), reindeer (*this page, bottom*), bison (*facing page, top*) and horses (*facing page, bottom*) collected along the banks of the Kolyma River, Duvanniy Yar, Siberia. All of the approximately 1,000 bones depicted here were collected in a single day and over an area of about 1 hectare. Photo credit: Sergey Zimov.

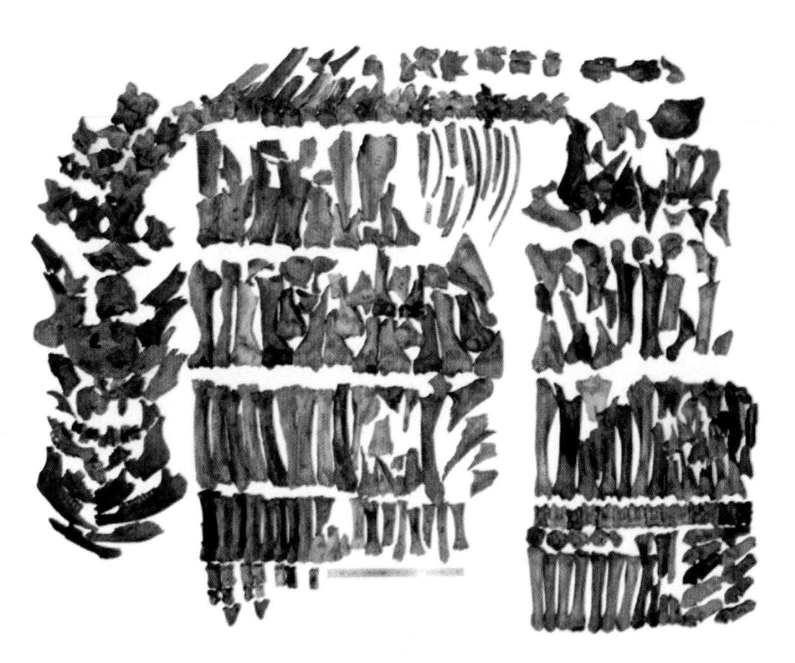

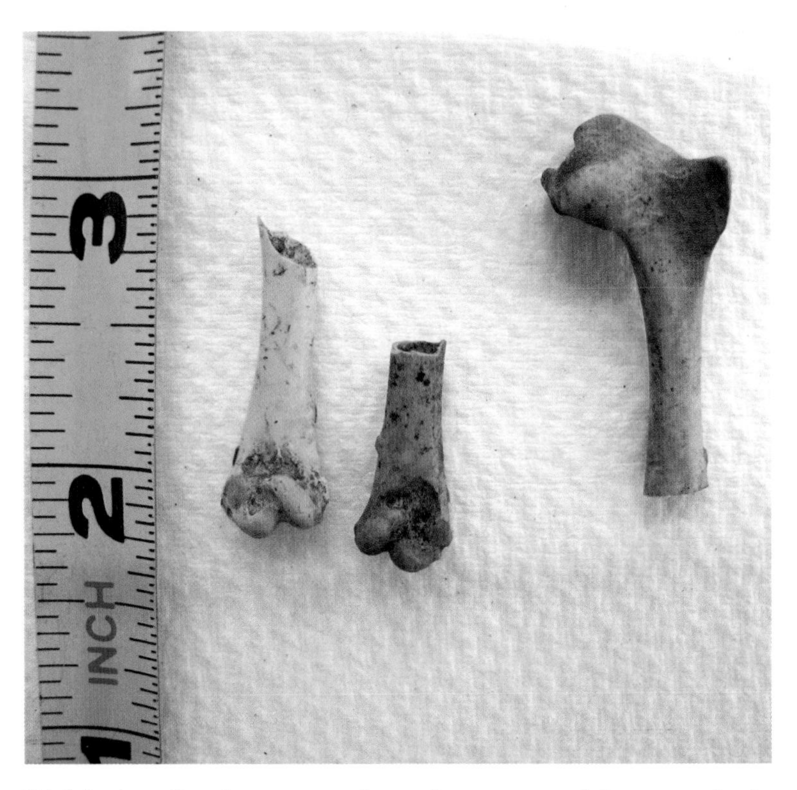

Plate 3. Leg bones from three passenger pigeons whose genomes are being sequenced at the University of California, Santa Cruz, as part of the passenger pigeon de-extinction project. These were among the remains excavated by Dr. Greg Sohrweide from a site in Onondaga, New York, USA, and date to the 1690s. Photo credit: Andre Elias Rodrigues Soares.

Plate 4. Sorting the remains of stingless bees from fragments of ancient amber in the ancient DNA facility at the Pennsylvania State University. Although amber-preserved insects were once thought to harbor preserved ancient DNA, research has shown that DNA does not survive in amber, even over relatively short periods of time. Photo credit: Mathias Stiller.

Plate 5. Field sampling of ice age bones. Only a small amount of tissue is required for DNA extraction and analysis. Here, a small fragment of bone is removed from a sample collected on the Taimyr Peninsula, Siberia, during our 2008 field season. Photo credit: Beth Shapiro.

Plate 6. Placer mining near Dawson City, Yukon Territory, Canada. Here, gold miners blast the frozen soil with high-pressure water to expose the gold-bearing gravels beneath. As the soil is washed away, bones, teeth, tusks and other remains are revealed and can be collected. Photo credit: Tyler Kuhn and Mathias Stiller.

Plate 7. The partial skull of an ice age horse recovered from an active placer mine near Dawson City, Yukon Territory, Canada. Photo credit: Tyler Kuhn and Mathias Stiller.

Plate 8. A cervical vertebral bone from a mammoth is slowly exposed by placer-mining activities near Dawson City, Yukon Territory, Canada. Sometimes, several bones from the same animal are recovered in close proximity. This particular mammoth bone was recovered in 2010 with four other vertebrae. Photo credit: Tyler Kuhn.

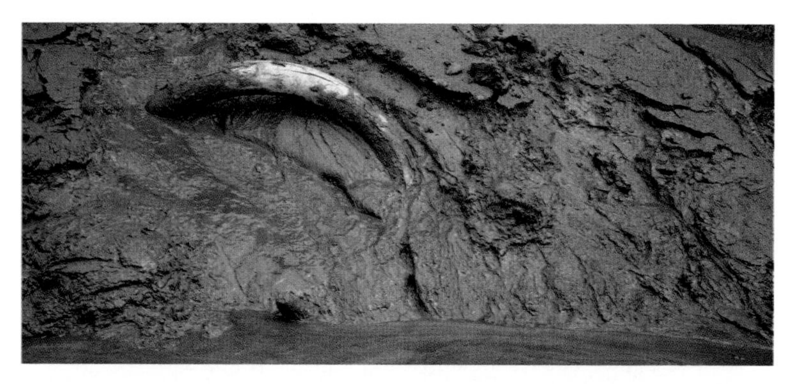

Plate 9. A mammoth tusk exposed by placer mining near Dawson City, Yukon Territory, Canada. Although it took several days for the soil surrounding the tusk to thaw completely, we eventually recovered the entire 2.5 m, 45 kg tusk. The tusk is now part of the paleontological collection of the Department of Tourism and Culture in Whitehorse, Yukon Territory, Canada. Photo credit: Tyler Kuhn.

Plate 10. As running water cuts through the permafrost, the remains of ice organisms are exposed. Geologists believe that a small stream began to cut through this area of permafrost near the Yana River in northeastern Siberia around 60 years ago. When the cut reached an ancient lake, rapid erosion formed what is now the Batagaika crater. Such fresh exposures are common along rivers throughout Beringia. Photo credit: Love Dalén.

Plate 11. A first look at base camp. Ian Barnes and I pose for a photograph as the helicopter is unloaded on the Taimyr Peninsula in the Russian Far North. Other members of the 2008 expedition team have already donned their mosquito-net hats. Photo credit: Beth Shapiro.

Plate 12. Setting up and settling in. As the mosquitoes swarm overhead, the 2008 Taimyr expedition team begins to stake out its tent sites. Our site is at the top of a hill surrounded by lakes, all of which we will search over the next weeks for the remains of mammoths and other ice age animals. Photo credit: Beth Shapiro.

Plate 13. Another shot from the first day of our 2008 Taimyr expedition: my tent, and several million mosquitoes. Photo credit: Beth Shapiro.

Plate 14. An ice cave beneath the city of Yakutsk, Sakha Republic, Russia. Caves such as this one are often used in Siberian cities to store food during the summer months. At the far end of this ice cave, scientists prepare to display the Yukagir mammoth to a delegation of international scientists attending a conference. Photo credit: Beth Shapiro.

Plate 15. Wild Spanish ibex escape manipulation by the scientists leading the bucardo-cloning project. Accustomed to climbing vertical rock faces and balancing on narrow ledges, the wild ibex easily balance atop a thin ledge within the captive breeding facility, well out of reach of the research team. Photo credit: Alberto Fernández-Arias.

Plate 16. Grazed and ungrazed land in Sergey Zimov's "Pleistocene Park" in the spring, after snowmelt. Ten years earlier, the area was a continuous community of willow shrubs. Today, the grazed area (*foreground*) in early spring has small amounts of green grass and freshly churned soil. This is caused by herbivores returning to this site during winter to graze and, in the process, trampling the snow and exposing the soil to the cold winter air. Photo credit: Sergey Zimov.

IF IT LOOKS LIKE A MAMMOTH AND ACTS LIKE A MAMMOTH, IS IT A MAMMOTH?

Let's return to the work that is going on in the present day. It is feasible today to use genome-engineering technologies to directly edit DNA sequences within a living cell. George Church's lab is using this technology to edit elephant cells so that the genome within them looks more mammoth-like than elephant-like. For now, this work is limited to editing only one or a few genes in somatic cells. However, we have somatic cells that contain genomes in which some genes have had the elephant version removed and replaced with the mammoth version. This is the status quo of mammoth de-extinction.

If the somatic cells edited by the mammoth revivalists are used to create a baby elephant, that baby elephant would have only a very tiny amount of mammoth DNA. The mammoth revivalists' goal is to engineer an elephant so that it can survive better in the cold. Let's imagine that they achieve this by replacing the elephant version of something in the range of five to ten genes with the mammoth version of those genes. In this scenario, the phenotype of the hypothetical baby elephant hopefully would change, but more than 99.99 percent of its DNA would still be elephant DNA.

In the fictional scenario published in 1984, the babies that were born were first-generation hybrids, created by fusion of DNA preserved in a mammoth egg and DNA from elephant sperm. The hybrid creature's DNA was 50 percent elephant and 50 percent mammoth, but ben-Aaron never went so far as to call them mammoths. In fact, she provided an entirely new scientific name—*Elaphas pseudotherias*—which places the hybrid mammontelephase in the same genus as the Asian elephant, but gives it an entirely new, and fictional, species name. Perhaps her goal was to be scientifically precise about what she created. Perhaps it was to avoid confusion. Whatever her motivation, the piece provides an excellent opportunity to observe the public's reaction to the creation of a (fabricated) hybrid species.

The public did not care that it was a hybrid. The media called it a mammoth, and so it was a mammoth. Perhaps what was most important was how it was described, but even this was absolutely minimal in the media reports: the hybrid had yellow-brown hair and mammoth-like jaws. Clearly, even a tiny bit mammoth-like was good enough for the public. It was a mammoth.

This is great news for those in favor of de-extinction, because it provides an enormous amount of wiggle room for determining when de-extinction is a success. A mammoth will not have to be pure in order to be received as a mammoth. This is a relief, because—as we've discussed—while 100 percent mammoth is out of the question, 1 percent mammoth may not be.

This provides an opportunity to redefine *de-extinction,* shifting away from a species-centric view. Genetically pure mammoths, or genetically pure versions of any extinct species, are likely not possible. However, we do not need genetic purity to benefit from de-extinction technology. If we select wisely which 1 percent of the genome to change, we may be able to resurrect those characteristics that distinguish a mammoth from an elephant. More importantly, we may be able to resurrect those characteristics that allow the elephant to live where the mammoth once lived. Once released into the wild, the hybrid elephant could stomp around, knocking down shrubs and consuming vast quantities of vegetation. It could disperse seeds and insects and distribute nutrients. This new hybrid animal could replicate the mammoth, without necessarily being a mammoth, with vast potential benefits to the arctic ecosystem.

Most people who are seriously considering either de-extinction or back-breeding are doing so because they believe that bringing these species back would provide an upper hand in present-day struggles to preserve biodiversity and maintain healthy ecosystems. Extinctions at any level—whether of prey species or predator species or species that distributed seeds or species that consumed shrubs and trees so as to preserve open spaces—can have cascading effects across an entire ecosystem.

The project to breed back the auroch in mainland Europe

aims to create giant herbivores that will graze wild, open land and thereby keep the shrubs and trees at bay. The result, the team hopes, will be a restored habitat that can be used by both large and small mammals and at the same time increase the diversity of plant species on the landscape. The auroch is the target phenotype of their back-breeding experiments. However, the team's intention is not to bring an auroch back to life but to resurrect a phenotype that can do in that environment what the auroch used to do. They hope to replace the auroch with something similar in function but not necessarily identical in form.

In my mind, it is this *ecological resurrection*, and not *species resurrection*, that is the real value of de-extinction. We should think of de-extinction not in terms of *which life form* we will bring back, but *what ecological interactions* we would like to see restored. We should ask what is missing from the existing ecosystem that could be recovered. De-extinction is perhaps better imagined as an elaborate bioengineering project in which the biological end product is modeled on something that evolution created but that has unfortunately been lost.

WHAT PARTS OF THE GENOME SHOULD WE ENGINEER?

Genome engineering, and not cloning by nuclear transfer or back-breeding, seems to be the most likely avenue to resurrect extinct traits and—depending on how loosely we care to define a species—extinct species. But where do we begin? The answer to this question is likely to be different for each de-extinction project.

If our goal is to create an elephant that is capable of surviving a Siberian winter, then we have to change this tropically adapted species into something that fares well in the bitter cold. Longer, thicker hair will definitely help, as will hemoglobin with higher efficiency in carrying oxygen at low temperatures. But what other traits should we try to engineer? Are there other ways to make an elephant more efficient at maintaining its internal body temperature? Are there energetic requirements to living in the Arctic that we haven't yet considered? Are there adaptations to the digestive system that will be necessary to allow an elephant to eat the plants that grow in Siberia? Do we need to engineer morphological changes that make the elephant capable of digging plants out of the snow? Will we need to engineer the elephant's immune system so that it can evade pathogens that are not present in the tropics? These are all good questions to which we do not yet have any answers, much less a target gene or suite of genes that we could sequence and look for mammoth-specific changes that we will want to engineer.

The scientific world is unlikely to prioritize elephant genomics in the near future, which means that we won't know anytime soon where all the genes are in the elephant genome, what these genes do, or how they interact with each other. This information, however, is all crucial if we really want to genetically engineer a mammoth in a piecemeal fashion. Given that so much is unknown, one solution might be to change *every* nucleotide in the genome where the mammoth differs from the elephant. In doing so, we would be less apt to overlook any important difference or interaction between genes. This would, however, require making a lot of changes: if we assume that the Asian elephant and the mammoth diverged from their common ancestor around four million years ago and that the rate of divergence is similar to that in other mammals, we can expect something like 70 million genetic differences between the two species (on the same order as the number of genetic differences that separate humans and chimpanzees). Less than 2 percent of the elephant genome would need to be edited, but 70 million changes is *a lot* of changes to make.

So how would we make those changes? First, we have to figure out what they are. Many (if not most) of the differences between the Asian elephant genomes and mammoth genomes can probably be identified by sequencing and assembling both genomes, lining them up, and scanning them for sites at which they differ from each other. Since we know that we will not be able to sequence and assemble a complete mammoth genome, we've already stumbled upon the first problem with this approach. Ignoring that problem, the next step is to design a strategy to change each of the elephant sites that differ into the mammoth version using genome-editing tools. If we assume that each edit will require its own CRISPR-RNA (the CRISPR-RNA is the part of the CRISPR/Cas9 system that finds and then binds to the part of the genome where the edit is to be made), then we need to design and deliver into the cell 70 million different CRISPR-RNAs. However, George Church's lab has been improving techniques to insert larger and larger fragments of DNA at once, which may allow us to change multiple bases at the same time. Let's assume that the technology gets really good, and we can make, on average, ten changes with each CRISPR-RNA. This would reduce the number of CRISPR-RNAs necessary to around 7 million.

In their mammoth hemoglobin work, George Church's revivalists designed two CRISPR-RNAs to make three changes to the hemoglobin gene (one CRISPR-RNA made a single edit and the other made two edits). Editing the elephant sequence takes place in three steps: First, everything necessary to edit the genome—the CRISPR-RNAs, Cas9 (the molecular scissors), and the mammoth template DNA—has to be delivered into the cell. Second, the CRISPR-RNAs have to find the part of the genome that they are intended to cut. Third, the cellular-repair machinery has to paste in the mammoth version of the gene.

Because the mammoth revivalists have actually performed this experiment, we can use their results to estimate the overall efficiency of the cut-and-paste process. In other words, we can ask, what proportion of edited elephant cells end up with all three changes? The mammoth revivalists found that each CRISPR-RNA had a different efficiency in targeting the right part of the genome (the "cut" step), and that the cellular machinery had a different efficiency in fixing each cut the way we want it to be fixed (the "paste" step). In this experiment, they estimated that the cut-and-paste efficiency of one of their CRISPR-RNAs was about 35 percent, and the other (the one that makes two changes) was about 23 percent. This means that only 8 percent of cells ended up with all three changes.

Even if we were able to reduce the number of CRISPR-RNAs that we need to make to, say, 100 (many fewer than the 7 or 70 million estimated above), and we assume, generously, that the efficiency of each of these is somewhere around 30 percent, that would mean that we would have to change at least 5×10^{53} cells in order to find one cell in which all 100 changes were made at the same time. That's a big number. To put this in some perspective (although perspective is very hard at this scale), scientists have estimated that there are around 40 trillion (4×10^{13}) cells in a human body and 7.5×10^{18} grains of sand on Earth.

Fortunately, we may be able to narrow down the number of changes we need to make without resorting to targeting specific traits. First, some of the species-specific differences that we observe when we compare one Asian elephant genome and one mammoth genome will not exist if we were to compare all mammoths and all elephants. These sites will look at first like they differ between species because we have only a single individual of each species to compare. But, if we were to have multiple genomes from each species, we would notice that some differences are not *fixed* within either species but are instead variable among mammoths or among elephants. Since not every mammoth or not every elephant has these changes, we can conclude that these changes are not important in making a mammoth look and act like a mammoth (or making an elephant look and act like an elephant). We could therefore exclude these sites from genome editing.

Another way to limit the number of necessary edits is to make only those changes that occur within genes. The genome is a big place, and only a small portion of the genome—around 1.5 percent of the human genome, for example—is made up of genes that code for proteins, while the rest of the genome is made up of other, noncoding DNA. Because genes code for proteins and proteins make phenotypes, the most important genetic differences between two species might be those that are found within the sequences of the genes themselves.

There are, unsurprisingly, several problems with this strategy. We do not, for example, know the location of all of the genes in the mammoth genome, so finding them will require educated guesswork—comparison with better-studied genomes—and even then we may not find all of the genes. Also, targeting only those differences that occur within genes may miss important differences that are found in the noncoding portion of the genome, such as differences that change when or how much of a gene is expressed. Differences in gene expression can result in different phenotypes even if the sequence of the gene itself is exactly the same.

Perhaps, then, we will need to make every change in the genome sequence. George Church believes that this will soon be feasible. The key, according to George, is to reduce the number of CRISPR-RNAs by cutting and pasting very long—very, very long—fragments of DNA. Instead of making a only few changes with each CRISPR-RNA, we will need to make thousands of changes, if not tens of thousands of changes, at once. Right now, George's group is able to synthesize strands of DNA that are 50,000 base-pairs long. While the accuracy of such long synthetic sequences is still less than ideal, the technology is improving while the cost is going down. If it were possible to synthesize the entire mammoth genome in, say, 100,000 base-pair chunks, then we could cut-and-paste the entire mammoth genome into an Asian elephant genome using fewer than 350 CRISPR-RNAs.

Still, 350 is a big number and, following the logic above, would require an absurd number of cells even if each cut-and-paste experiment worked with exceptionally high efficiency. The logic presented above is not particularly logical, however, and does not describe how we would perform the experiment in reality. Rather than try to luck into a scenario in which 100 (or 350) things that are unlikely to happen all occur at the same time, we would perform the experiment in steps, where a few changes will be made and validated, and then a few more introduced to those cells that were edited successfully, and so on. The experiment would still be challenging, and it would still take a long time to complete, but it might just be feasible.

Today, we do not know the complete genome sequence of the mammoth. However, we are likely to learn most of the mammoth genome sequence within the next few years. Today, we cannot edit an Asian elephant genome so that it looks entirely like a mammoth genome. This technology is also improving. In fact, this particular step in the de-extinction process is probably the fastest moving in terms of technology development.

MORE THAN THE SUM OF ITS NUCLEOTIDES

Genome editing will become an increasingly efficient way to transform all or part of a genome of a living species into something that resembles the genome of a species that is extinct. However, some important differences between species may have nothing at all to do with the sequence of their genomes. Simply changing the genome sequence might not, therefore, be sufficient to resurrect the extinct phenotype.

Genomes are complicated places. Genomes live in cells, which live in bodies, which live in environments. And in different cells, different bodies, and different environments, the same genomes—genomes that are identical in both the coding and noncoding portions—can produce very different phenotypes. Identical twins, for example, have identical genomes. But, as they get older, identical twins become more and more phenotypically and behaviorally different from each other. How can this happen, if their genomes are the same?

In addition to their genome, all organisms have what is called an *epigenome*. The epigenome is a confusing concept, and not all scientists define or describe the epigenome in the same way. As I understand it, the epigenome can be thought of as a suite of tags that that are attached to the genome. These tags indicate whether a gene is turned on (making proteins) or off (not making proteins). Importantly, these tags are not actually *part of* the genome, which means they can be in flux throughout the organism's life. Epigenetic tags can be heritable—that is, the epigenetic state of a particular gene is sometimes passed on from parent to offspring. These epigenetic tags might tell a cell to turn on only those genes necessary for being a heart cell, for example. Other epigenetic tags are not heritable in this traditional sense and, instead, may appear or change because of interactions that take place between the organism and the environment in which that organism lives.

A variety of environmental stimuli are known to affect the epigenome. An organism's diet, the amount of stress or toxins it is exposed to, and how much physical exercise it gets will all alter the epigenome, changing which genes are expressed, when they are expressed, and how much they are expressed. By the time identical twins become adults, their epigenomes differ considerably, although their genomes remain identical. It is the combination of their genome sequence and the epigenetic differences that accumulate over each twin's lifetime that results in their distinct phenotypes.

Will epigenetics complicate de-extinction efforts? We don't know. If we edit an elephant gene to contain mammoth DNA sequences, it will, as it begins to develop, contain the elephant epigenome. In the womb, it will be exposed to the developmental environment of an elephant: a mom that eats an elephant diet, lives in the elephant habitat, and expresses elephant genes. It will survive by virtue of the elephant placenta, which will be expressing elephant genes modified by that particular mother elephant's epigenome.

While we cannot study the effects of the developmental environment using identical twins (because they develop in the same intrauterine environment), we know the health and diet of the mother during pregnancy can have profound effects on fetal development. Her diet can even affect health outcomes later in life, such as risk of heart disease and obesity. Fascinatingly, we also know that the mother's diet *before* pregnancy can influence the epigenetic state of her genes, with consequences to the developing embryo. Almost certainly, the diet and amount of stress to which the mother elephant is exposed will affect her developing mammoth (or mammoth-like) embryo, but exactly what these effects will be remains unknown.

In some instances, a species-specific developmental environment is not critical to a successful gestation. Robert Lanza's genetic-engineering firm, Advanced Cell Technology, successfully cloned both a gaur and a banteng (both living but endangered species that are closely related to cattle) using nuclear transfer and female domestic cows as surrogate mothers. Both pregnancies went well, and both calves appeared to thrive. It is unclear, however, how these animals might differ from clones that were born from surrogate mothers of their own species.

What about the environment *after* birth? Epigenetic changes accumulate throughout life and are driven by the environment in which an organism lives. How much of looking and acting like a mammoth is due to having a mammoth genome, and how much of it is due to living life in the steppe tundra? This is something we may have to wait to learn.

Understanding the genome and how the genome interacts with the environment is among the major technical hurdles standing in the way of successful de-extinction. It is unclear today how this hurdle will be surmounted. Will we finish sequencing the mammoth genome and learn where all of the genes are and what all of the genes do, so that we can make a minimal number of changes and still end up with a mammoth? Or will genome-editing technology advance to the point where we can make all the changes necessary to create a genome that is 100 percent mammoth-like? Will we devise a way to infer the epigenetic state of ancient tissues, as a first approach to learning which genes should be turned on or off in unextinct individuals?

Answers to these questions may come soon. Knock-in and knock-out experiments—where scientists either turn on or turn off specific genes in organisms like yeast, fruit flies, and mice—are being used to discover where genes are, what they do, and how they interact with each other. Large, population-level human genome-sequencing projects are being used to identify specific genetic changes that are associated with distinct phenotypes, such as adaptation to life at high altitudes or susceptibility to cancers or other diseases. These experiments are homing in on ways to identify what are likely to be the most "important" changes to make. At the same time, the technology behind CRISPR/Cas9 systems is developing rapidly. These systems have so far been used to edit the genomes of more than twenty different species, chopping out and inserting fragments of the genome that are on the order of tens of thousands of nucleotides long. We may eventually arrive at a solution where it is possible to edit an entire genome.

Ancient epigenomes may even be within reach, thanks in part to how DNA degrades over time. It turns out that DNA methylation, which is one way that the epigenome marks the genome, interacts with DNA degradation in an interesting and useful way. In methylation, the epigenome modifies the genome by attaching a methyl group (CH_3) to a cytosine—one of the four nucleotide bases that make up DNA. DNA degradation also affects cytosine bases, but in a different way. Cytosine bases are often deaminated as DNA degrades—they lose part of their chemical structure (an amine group) and become uracil, which is a nucleotide base that is otherwise not found in DNA. When methylated cytosine bases become deaminated, however, the interaction between the two chemical modifications converts the cytosine into thymine, another of the four nucleotides found in DNA, rather than uracil. The ancient epigenome can be reconstructed by distinguishing deaminated cytosine bases that become thymine bases (which degraded after being tagged by the epigenome) from those that become uracil bases (which also degraded but were not epigenetically tagged).

This approach was first used by Ludovic Orlando's research group at the University of Copenhagen in Denmark to reconstruct the epigenome of a 4,000-year-old Paleo-Eskimo from Greenland's Saqqaq culture. Soon afterward, a team of scientists from the Max Planck Institute for Evolutionary Anthropology in Leipzig, Germany, and the Hebrew University of Jerusalem, Israel, mapped the epigenome of two archaic hominins—a Neandertal and a Denisovan. The team found around 2,000 differences between the reconstructed epigenomes of the archaic hominins and the epigenomes of modern humans, some of which may underlie some of the skeletal differences between us and our archaic cousins.

While technologies to sequence, edit, and understand genomes are all developing at a rapid pace, new tools that become available tend to work best for those species that are the best studied. Far less is understood about elephants than about mice, fruit flies, or humans, and the same is true for many of the candidate species for de-extinction. These tools can be adapted for research on other species, but, for now, the hurdles standing in the way of fully reconstructing the genomes of extinct species remain high. George Church, however, is very tall.

CHAPTER 8

NOW CREATE A CLONE

I have, up until this point, been very clear that mammoths will not be resur-rected by cloning. What I say next, therefore, is likely to be confusing. *The next step in bringing a mammoth back to life is to create a clone.*

In my defense, the cells that we would clone at this stage would be very different from the cells that the Japanese and South Korean teams hope to find and use in their cloning experiments. By the time we arrive at this stage of de-extinction, we might have spent years (even decades) in the lab, painstakingly engineering changes to the elephant genome within our cells. We would not be beginning our cloning experiments with miraculously well-preserved mammoth cells. Nonetheless, the next step in de-extinction would be to "clone" our cells, and thereby turn them into an entire elephant (with some mammoth genes).

Some de-extinction projects will, of course, be able to skip the genome-engineering steps and proceed directly to cloning. These projects may move ahead much more quickly than those that require genome engineering. Of course, that simply means that they will be first to arrive at the next hurdle. Consider the example of the bucardo.

NOT QUITE THE FIRST DE-EXTINCTION

In the summer of 2003, a young female bucardo, which is a subspecies of Spanish ibex (a type of wild goat), was born. Bucardos had been endemic to the Pyrenees, the mountain range that forms the border between Spain and France. When this baby bucardo was born, however, bucardos had been extinct for just over three and a half years.

The baby bucardo was a genetic clone of the last living bucardo, an elderly female named Celia. Unfortunately, the baby suffocated within minutes of birth. An autopsy revealed that she had been born with a malformed lung and had no chance of survival. Nonetheless, the birth of this baby bucardo is often held up as the first successful de-extinction. I disagree. To me, if she had *no chance of survival,* this is not de-extinction.

As de-extinction projects go, the bucardo project has a lot of promise. Bucardo cells were harvested from Celia ten months prior to her death and immediately frozen, and the DNA within those cells is in very good condition. Several closely related subspecies of Spanish ibex are still thriving, so finding appropriate egg donors or surrogate mothers should be straightforward. The bucardo has also not been extinct for very long, and its extinction was likely due to overhunting and not to the disappearance of its habitat. As long as we can control our guns, resurrected bucardos could be returned to their native habitat without the need for extensive environmental impact studies or political maneuvering.

When the team of Spanish and French scientists began the bucardo project in 1989, bucardos were not yet extinct. Cross-species cloning had also not yet been achieved in large mammals, and the challenges facing the project were immense.

In 2001, and in a separate effort to perform cross-species cloning, the biotechnology company Advanced Cell Technologies successfully cloned a gaur, which is an engendered species of cattle native to South and Southeast Asia, using a cow as a surrogate host. The cloned gaur lived only forty-eight hours before succumbing to dysentery, but its birth demonstrated that cross-species cloning was possible. Two years later, the same company successfully cloned a banteng, another endangered cattle species from Southeast Asia, again using a cow as a surrogate host. This banteng lived in the San Diego Zoo for seven years—less than half its lifespan in the wild—before dying of what appeared to be natural causes.

The bucardo project was similar to the gaur and banteng projects in that there was no need for genome sequencing or genome engineering for cloning to be possible, and in that surrogate hosts were available. There were, however, two important differences that distinguished the bucardo project from the others. First, assisted reproductive technologies were already established for cattle, but had not yet been developed for Spanish ibex. Second, by the time the team of scientists developed this technology, the bucardo had gone extinct.

Unfortunately, the bucardo cloning experiment did not succeed, and it is not entirely clear why. It is possible that the experiment failed because the scientists simply did not make enough embryos. Cloning by nuclear transfer is, after all, notoriously inefficient. The team transferred copies of Celia's somatic cells into 782 eggs, but only 407 eggs developed into embryos. Of these, 208 embryos were transferred into potential surrogate hosts, but only seven pregnancies were established. Only one pregnancy lasted to term, and the baby bucardo that was born lived for less than ten minutes. If one were to count this baby bucardo as a successfully established clone, which I will do here only for the purpose of illustration, the success rate of bucardo cloning would be 0.1 percent.

Alberto Fernández-Arias, who is the director of the Aragón Hunting, Fishing, and Wildlife Service and who was brought in to the bucardo project in 1989 to develop assisted reproductive technologies in Spanish ibex, believes that I am being unfair in my characterization of this as a "failed" de-extinction. He points out that if his team had known that the bucardo would be born with a lung deformity, they could have been prepared to remove the malformed part of the lung immediately after birth. Such surgery has been performed successfully on human babies with similar birth defects and, indeed, may have been able to save the baby bucardo's life. Of course, it is not possible to know either what caused the lung deformity or what might have happened—how the bucardo would have developed, how it might have fared in adulthood—if the bucardo had survived. The project continues, however, and we may soon have another opportunity to find out whether bucardos will once again roam the Pyrenees.

De-extinction by Nuclear Transfer

Once we have a cell that contains the genome of the animal we want to create, whether that cell has been grown from frozen tissues that were harvested prior to extinction or subjected to genome editing, the next step is to create an embryo from that cell. This involves using a living host as a surrogate. In many candidate species for de-extinction (I will describe some exceptions later in this chapter), this involves cloning by nuclear transfer. And, as one might anticipate, some candidate species for de-extinction will be considerably easier to clone than others will be. Cloning the bucardo, for example, *should* be much simpler than cloning edited elephant cells would be. For that reason, I will begin the exploration of this phase of de-extinction using the bucardo as an example. Once the basics have been covered, I will move on to the bigger challenges to be faced when cloning genetically engineered elephant cells. And, finally, I will introduce an obstacle to de-extinction that took me completely by surprise: it is not possible to clone birds.

The Making of a Bucardo

Nuclear transfer is a complicated process with potential disaster lurking in each step. Even what should be the simplest steps can harbor significant obstacles. With dogs, for example, it is nearly impossible to harvest mature eggs—that is, the eggs into which the somatic cell would be transferred—from female dogs. Unlike the eggs of other mammals, which mature in the ovary, dog eggs mature as they move from the ovary into the uterus. Because domestic dogs also tend to have unpredictable ovulation cycles, knowing when to harvest mature eggs requires both careful monitoring of the dog's hormones and a bit of luck.

The most challenging step in nuclear transfer is, however, reprogramming. During reprogramming, the cell forgets how to be a somatic cell and becomes, essentially, an embryonic stem cell. Only cells that have reset completely can later differentiate into all of the various cell types that make up an organism. This step is, however, particularly inefficient. Incomplete reprogramming is thought to explain why so few embryos develop after nuclear transfer and the high frequency with which developmental defects are observed among those embryos that do develop.

Reprogramming is not the only step that can fail. Even if cells are reprogrammed correctly and develop into viable embryos without developmental defects, the egg may fail to implant in the uterus of the surrogate mom, or the pregnancy may fail after implantation. This could be due to a poor understanding of the reproductive cycle or to some kind of incompatibility between the developing embryo and the surrogate mother. Such incompatibilities are likely to be more common among cross-species clones (including de-extinction experiments, where all or part of the genome is from a different species than the surrogate mother) compared with same-species clones. Also, there is little doubt that experimental manipulation is stressful for the surrogates and that this stress may contribute to the elevated rate of failed pregnancies in cloning experiments.

Anxious Ibexes and a Hybrid Solution

Stress was certainly a limiting factor in the bucardo cloning experiments.

In preparation for working with bucardo cells, the scientific team leading this work first attempted a cross-species cloning experiment with a different and relatively common subspecies of Spanish ibex. Once these technologies had been developed and fully tested, the team would proceed with the bucardo.

The plan required Spanish ibex embryos. To create these embryos, the scientists first had to capture Spanish ibex from the mountains. Then, they needed to rear the captured ibex in captivity to observe their reproductive behavior and develop a way to make the females ovulate. After mating had been observed, the scientists would harvest fertilized eggs, implant the developing embryos into domestic goats, and hope for the best.

Harvesting Spanish ibex eggs turned out to be much more difficult than the team anticipated. Accustomed to climbing the steep faces of rocky slopes, the Spanish ibex escaped manipulation by taking refuge along high ledges in the walls of the animal facility (plate 15). When the team finally harvested their eggs, the eggs were all unfertilized. The animals, it seemed, were too stressed by the captive breeding environment to mate successfully.

The team was able to develop less stressful ways of manipulating the ibex and, eventually, they recovered fertilized eggs from captive Spanish ibex. Any excitement derived from this success was short-lived, however, as another serious setback to the experiment was revealed: none of the embryos continued to develop after implantation in domestic goats. It seemed that the domestic goat uterus was incompatible with the Spanish ibex embryo. This was bad news for bucardo cloning.

Believing that genetics might be the solution, the team decided that a different surrogate mother—one that was genetically more similar to the developing embryo—might be just what they needed. The most genetically similar surrogate would be a subspecies of Spanish ibex. They knew, however, that Spanish ibex were difficult to manipulate and did not fare well in captivity. Not wishing to spend every day coaxing ibex down from the walls, they decided on a compromise: they would create hybrids. Female domestic goats crossed with male Spanish ibex would produce kids with 50 percent Spanish ibex DNA and, most crucially, that would probably keep their feet on the ground. When these hybrid females reached adulthood, they would become the surrogate mothers for the Spanish ibex embryos.

Around a year later, the team transferred Spanish ibex embryos into hybrid goat-ibex females and, again, hoped for the best. Excitingly, half of the embryos established successful pregnancies and developed into healthy Spanish ibex.

I should point out that this success rate—50 percent survival of implanted embryos—is high because this particular experiment did not involve nuclear transfer. This experiment began with healthy embryos taken from living ibex, not with somatic cells that needed to be reprogrammed. As I noted before, this reprogramming step—which is the first step in bucardo de-extinction—has an extremely low success rate.

Unanticipated Barriers to De-extinction

In developing assisted reproduction technology for Spanish ibex, the bucardo-cloning team learned that bucardo embryos, should they get that far in the bucardo-cloning experiment, might develop within surrogate mothers that were hybrids of domestic goats and Spanish ibex, but they were unlikely to develop within purebred domestic goats. The team had discovered a barrier of some sort to cross-species cloning that had arisen during the evolutionary divergence between these two lineages.

Importantly for de-extinction, the probability that barriers such as this may arise increases with evolutionary distance. Extinct species with no close evolutionary relatives might not have any suitable living potential maternal host. The ibex experiment revealed, however, that barriers can also exist between closely related species. Genome editing could even *cause* barriers—for example, by disrupting important interactions between the embryo and the maternal host. In this way, even de-extinction projects that involve only minimally edited genomes may be frustrated by unanticipated incompatibilities between the developing embryo and the surrogate host.

Some incompatibilities may manifest even before the implantation phase. If, for example, the egg cell into which the nucleus of the somatic cell is injected is incompatible with the somatic cell, then none of these eggs will develop into embryos even if the somatic cell is completely and correctly reprogrammed. Such a problem may arise, for example, if the nuclear genome from the somatic cell is incompatible with the mitochondrial genome in the egg cell.

Mitochondria are organelles that live within the cytoplasm of cells and are not part of the nuclear genome. All of the mitochondria that an organism will have in all of its cells are descended from the mitochondria in the egg cell. Mitochondria have their own genome, and this genome codes for some of the proteins that are necessary for cellular respiration—that is, the process by which cells use oxygen and sugars to make energy. Other proteins necessary for cellular respiration are made by genes in the nuclear genome. Incompatibility between the mitochondrial and nuclear genomes can lead to incompatibilities between these genes. If these genes can't work together to make the cell respire, this can lead to metabolic disease, neurologic disease, and even death. Thus far, all cross-species cloning has involved the transfer of only nuclear DNA—*not* mitochondrial DNA.

Researchers in David Rand's lab at Brown University demonstrated how nuclear-mitochondrial mismatch can produce unusual phenotypes in otherwise genetically normal cross-species hybrids. Rand's lab created fruit flies with nuclear DNA from *Drosophila melanogaster* and mitochondrial DNA from *Drosophila simulans*, two fly species that diverged from each other around 5.4 million years ago. The resulting mismatched-genome flies had whiskery bristles on their backs, were half the length of normal flies, were developmentally delayed, reproduced poorly, and, as might be expected if energy production is off, got tired more quickly than did flies with matched genomes.

Mismatches between the mitochondrial and nuclear genomes might be a problem for de-extinction, but not one without an obvious solution. If the mitochondria don't match, why not replace them with mitochondria that do match the nuclear genome? Or, why not edit the mitochondrial genome to replace the problematic sites? This could presumably be achieved using the same genome-editing approaches as would be used to alter the sequence of the nuclear genome. Neither of these approaches is simple, and neither is feasible today. Both are, however, theoretically possible.

MAMMOTH PROBLEMS

Now that I've introduced some of the challenges to be faced during the cloning and prenatal development stage of de-extinction, let's return to the mammoth as a specific example. As I discussed in the preceding chapter, we have the technology today to edit the elephant genome so that it contains the mammoth version of least some genes. Assuming that genome engineering takes place in a cell that either is a stem cell or can be reprogrammed to become a stem cell, we are ready to move on to the next step: creating a living animal that contains the edited genome and, hopefully, expresses the traits that we aim to resurrect.

To complete this step, the cell needs to develop into an embryo, and because we cannot grow an elephant in the lab, that embryo needs to be transferred into a surrogate host. Once inside the surrogate, the embryo needs to implant in the uterine wall and establish a pregnancy. The pregnancy then needs to proceed without problems and culminate in the birth of a healthy baby animal whose genome contains several carefully selected and painstakingly engineered mammoth genes.

The simplest way to transform the edited cell into an embryo is to use an egg. We know that eggs contain proteins that activate cells—that is, they reset cells that have already differentiated and turn them into embryonic stem cells. The most appropriate egg to activate our edited elephant cell is, unsurprisingly, an elephant egg. Elephant eggs are not particularly easy to come by. When an Asian elephant ovulates, she releases only one egg at a time. Once released, the egg travels through her reproductive system to the uterus, which is, predictably, elephant sized. An elephant that is not pregnant will ovulate once every two to three months. Given the poor efficiency of nuclear transfer, it's reasonable to assume that collecting a single egg every two months—assuming we can find that egg within the elephant's reproductive tract—will not provide enough eggs. We'll need hundreds, even thousands of elephant eggs for this to work. Frankly, that seems unfair. Elephants are struggling to make enough elephants to sustain healthy populations; the last thing they need is for us to be snooping around their ovaries stealing their precious few mature eggs. In fact, if harvesting mature eggs from adult elephants were the only way to get elephant eggs, my opinion would be that mammoth de-extinction research should stop *immediately*.

Fortunately, there may be another way. In 1998, researchers at Purdue University and the Advanced Fertility Institute at Methodist Hospital of Indianapolis created mice that could grow elephant eggs. Dr. John Crister, who led the study, wanted to develop a way to increase the reproductive rate of endangered species, and he hoped that coaxing laboratory mice to grow their eggs would be a good start. He and his team transplanted ovarian tissue—the tissue in which immature eggs are found—that Crister collected from three South African elephant carcasses into several laboratory mice. A few of these mice developed egg-producing follicles and, ten weeks later, one of these follicles produced a slightly misshapen elephant egg. Crister and his team did not attempt to fertilize the egg with elephant sperm, so it's not possible to say whether it would have developed into a viable embryo. It is, however, an optimistic start.

Hopefully, scientists will invent an efficient means to collect a large number of elephant eggs without jeopardizing any elephants. We could then collect a ton (perhaps literally) of elephant eggs, remove their nuclei, and insert nuclei that contain our edited genomes. Then, we would sit back and allow the egg to perform its reprogramming magic. If this goes smoothly and we end up with viable, developing elephant embryos (with slightly modified genomes), we can then transfer these embryos into the uteri of adult female elephants, where they can develop into baby elephants (with slightly modified genomes).

The entrance to an elephant's uterus is blocked by a membrane called a hymen. In elephants, the hymen stays in place throughout pregnancy, ruptures during birth, and then grows back in preparation for the next pregnancy. To establish a healthy pregnancy in a surrogate elephant mother, the embryo and whatever tool is used to deliver it into the uterus must pass through the only opening in the hymen—a four-millimeter hole designed to allow only sperm to penetrate—without destroying the membrane and thereby compromising the pregnancy.

Let us assume this is possible. Let's also assume that the pregnancy takes hold, and the embryo begins to develop. The next step is to wait patiently while the pregnancy proceeds. A typical gestation period for an Asian elephant is somewhere in the realm of eighteen to twenty-two months. Hopefully, no compatibility issues will develop between the embryo and the surrogate mother during this time. Hopefully, the surrogate mother's genetic makeup won't influence the expression of the genes we changed. Hopefully, her diet, hormones, and stress level won't alter the developmental environment in a way that influences the expression of the genes that we changed. Hopefully, the birth goes well for both the surrogate mother and the neonate.

Size Matters

In designing cross-species cloning experiments for the purpose of de-extinction, it is important to consider physical differences between the two species involved. Mammoths that lived during the Late Pleistocene varied considerably in size. The largest of these mammoths were about the same size as big African elephants and the smallest were similar in size or even smaller than average-sized Asian elephants. It is not known whether this size variation was genetically determined or simply reflected differences in the amount and quality of available food. Regardless, this variation might be important in choosing surrogate hosts. Interestingly, the two baby mammoth mummies that have been found were both around ninety centimeters tall, which is approximately the same size as a newborn Asian elephant, suggesting that the most closely related elephant species to a mammoth might make a reasonable surrogate.

Physical differences in size can cause problems in gestation and birth, however. Imagine, for example, if sperm from a Great Dane was used to impregnate a Chihuahua. The embryos would begin to develop and fill whatever space was available, but development would stall as they ran out of room to grow. Ultimately, the embryos may die, the mother may die, or both may die. If a natural birth were attempted, the mother would almost certainly suffer terribly. Returning to de-extinction, what might happen if a very large auroch were to develop within a much smaller, domestic cow? Or if a dugong tried to gestate a Steller's sea cow? Size differences between species, even between closely related species, should definitely be considered when proposing surrogate hosts.

A possible solution might be to make miniature versions of some extinct species. We could identify which genes or suite of genes are most critical to determining body size and tweak these using genome editing. A useful clue about which genes to target could come from genetic analysis of the population of mammoths that lived on California's Channel Islands. These so-called pygmy mammoths grew to only around two meters tall at the shoulder, compared with four meters or more for mainland mammoths, and probably weighed just under 800 kilograms, compared with 9,000 kilograms or more for mammoths on the mainland. There is one problem with this idea. Tiny mammoths may be easier to gestate, but they might not be sufficiently large to replicate the ecological interactions between normal-sized mammoths and the ecosystems in which these normal-sized mammoths lived. Resurrecting pygmy mammoths therefore might not achieve the environmental goals of mammoth de-extinction.

Another potential solution is to give up entirely on surrogates and instead use artificial wombs. I'm imagining something similar to the artificial wombs that Aldous Huxley envisaged to grow children in his book *Brave New World*. Or, even better, the giant nutrient-filled flasks in which human clones were grown on the planet Kamino to fight for the good side in the movie *Star Wars: Episode II*. In the artificial womb scenario, embryos would develop to term in a completely artificial environment—an idea known as ectogenesis. Modern medicine is a long way from functional artificial wombs and successful ectogenesis, but there is little doubt that innovations in this realm would have considerable impact on neonatal and perinatal care. Plus, by using artificial wombs, any animal suffering caused by surrogacy would be avoided entirely. The use of artificial wombs assumes, however, that developing within a real uterus is *not* critical to normal mammalian development. This is something that science does not yet know.

CLONING IS FOR THE BIRDS (NOT)

Although my focus until now has been on mammoth de-extinction, the present discussion provides an ideal opportunity to shift to the other de-extinction project with which I am involved—resurrecting the passenger pigeon. I hinted earlier that some species would not be cloned using nuclear transfer. The passenger pigeon is one of those species.

Because birds develop on the outside, rather than within the bodies of their surrogate moms, birds would seem to be a good choice for cloning by nuclear transfer. Yet, none of the species listed as having been cloned using this approach were birds. Why is that?

The simple answer is that birds cannot be cloned in this way.

A bird begins its long journey to becoming a bird as a yolk. The yolk is a single unfertilized cell—the oocyte—that lives inside the bird's ovary. The first step in bird development is to release the yolk into the oviduct. As it begins its journey down this very long, very spirally tube, it meets sperm and is fertilized. Then, for the next twenty-four hours or so, the fertilized egg travels slowly through the oviduct, tumbling around dramatic twists and through spiraled coils. As it bobs along its path, layers of albumin and structural fibers gradually cover the egg. This is the stuff we know as egg white. As it is moving, the fertilized cell starts to divide. The egg tumbles through the oviduct, twisting the structural fibers around the yolk, anchoring it firmly within the egg white. Toward the end of the oviduct, and just before the egg is laid, the hard shell is deposited as the final layer around the developing embryo. By the time it completes the journey from inside its mother's ovary to the outside world , the embryo comprises around 20,000 cells. These will have begun to differentiate into different cell types.

At what point in this process would it be possible to perform nuclear transfer? In a mammal, the egg whose nucleus is removed and then replaced is collected from the female reproductive tract after it has matured but before it is fertilized. At precisely this stage, the egg is primed to reprogram the nucleus of the somatic cell. It turns out to be extremely challenging to collect bird eggs that are at this stage of development. The reproductive tract in birds is long and sinuous, and the yolk is tricky to recover prior to fertilization. If we wait until the egg has been laid, the cells in the embryo will have already started to differentiate, and the embryo—which is held in position within the egg by many layers of twisted fibers—will be too large to remove. Even if the embryo could be removed and replaced without destroying the egg, the replacement embryo would have to be at the same developmental stage as the egg's natural embryo. Growing embryos to such a late stage in the lab is also proving to be extremely challenging. For the moment, therefore, it seems that cloning birds may never be possible.

Fortunately, there is another way. When the bird egg is laid, the embryo is still in an early developmental stage. The primordial germ cells—those cells that will later develop into either the sperm cells or the egg cells of the developing embryo—have formed but have not yet found their way to the sex organs, as the sex organs do not yet exist. Around twenty-four hours after the egg is laid, the primordial germ cells migrate through the developing embryo's bloodstream to the sex organs (which are now starting to develop), where they settle in until the time comes when they mature into sperm or eggs.

Primordial germ cells are the key to genetically manipulating birds. Primordial germ cells can be grown in a dish in the lab, which makes their genomes accessible for editing. Primordial germ cells are also tiny, which means they can be injected into the egg during that second twenty-four-hour window of development during which the egg is on the outside and the primordial germ cells are making their way to the bird's developing sex organs. In this way, the injected edited primordial germ cells will travel with the embryo's primordial germ cells to the sex organs.

When these cells mature, the edited cells will participate in making the next generation of birds.

When the chick hatches from the egg into which the genetically modified primordial germ cells were injected, that chick itself will not be genetically altered. Instead, the genetically altered cells will be hiding out in its sex organs. The first time the genetically altered genes will be expressed will be when *that* chick grows up and has its own baby chicks.

Let's walk through how this process would work for passenger pigeon de-extinction. Band-tailed pigeons are the closest living relative of passenger pigeons. The intention of the passenger pigeon de-extinction project, although these experiments have not yet begun, is to create band-tailed pigeons that look and act like passenger pigeons. To achieve this, we will isolate primordial germ cells from band-tailed pigeons and grow these in the lab. We will then edit the genomes within the primordial germ cells using the genome-engineering technologies described several chapters ago, replacing band-tailed pigeon genes with the passenger pigeon version of these genes as appropriate. We will then inject these edited band-tailed pigeon primordial germ cells into developing band-tailed pigeon eggs at precisely the right time during development. The chicks that are born when these eggs hatch will be genetically pure band-tailed pigeons, except that some of their germ cells (sperm or eggs) will contain passenger pigeon DNA. The offspring created by these edited germ cells will contain passenger pigeon DNA throughout their bodies.

Cloning by Germ Cell Transfer

Cloning by transferring germ cells into a developing embryo has one important advantage when compared with cloning by nuclear transfer. Edited primordial germ cells do not need to be reprogrammed. This is *huge*. So why has all the focus been on cloning a mammoth, when cloning passenger pigeons or dodos would apparently be so much simpler?

It is not entirely clear why cloning birds has received far less attention than cloning mammals has. Primordial germ cell transfer works remarkably well as a means to genetically modify birds. The technology has been developed mainly with the chicken industry in mind, but it has been used both for conservation purposes and in pure scientific research. There is no obvious reason to suspect it would not work well for the purposes of de-extinction.

Some of the applications of primordial germ cell transfer are, admittedly, unusual. The Roslin Institute, the facility that was responsible for cloning Dolly, has used the technology to create chickens that glow a bright green color under ultraviolet light. To make their chickens glow, they insert a protein into their genomes called *green fluorescent protein*, or GFP, which is found naturally in the North American jellyfish *Aequorea victoria*. The scientific community uses GFP to track biological changes within an organism. For example, if tissues whose cells express GFP are grafted onto an organism whose cells do not express GFP, scientists can track what happens to the grafted cells by watching them under fluorescent light. Scientists interested in using glowing chickens for their research can go to the Roslin Institute's Web site and order them. For now, there is no charge.

In addition to making chickens glow, the technique of injecting primordial germ cells into developing embryos has been used to boost the population sizes of rare or endangered chicken breeds. Primordial germ cells can be harvested from the blood of developing embryos without killing the embryo. These cells can then be kept alive in the lab and introduced into the developing embryos of common breeds. When these birds reach sexual maturity, they can then be fertilized with sperm (which are much easier to collect than eggs) from the rare breed. When these sperm fertilize eggs that develop from the injected primordial germ cells, the result is a purebred rare-breed chicken that hatches from an egg laid by a common chicken.

The most exciting application of primordial germ cell transfer from the perspective of bird de-extinction has been the successful transfer of primordial germ cells between species. Scientists at the Central Veterinary Research Laboratory in Dubai injected primordial germ cells from a chicken into duck eggs. When the ducks hatched from these eggs, they looked like perfect ducks. Remember, only the germ cells are different in the first generation. The scientists later harvested sperm from these ducks, and used those sperm to fertilize a hen. When the eggs laid by this hen hatched, perfect chickens were born. With a duck for a dad.

Fascinatingly, ducks and chickens are not the only animals that have been coaxed to give birth to a different species using this approach. Recently, Professor Goto Yoshizaki of the Tokyo University of Marine Science and Technology transferred rainbow trout eggs and sperm into the reproductive tracts of adult Masu salmon. When these adults mated, some of their eggs hatched into rainbow trout. Rainbow trout and salmon are closely related species, which may explain the experiment's success. However, there is hope that the technique can be extended to other fish species. Yoshizaki also produced tiger puffer fish using grass puffer fish and intends to use mackerel to produce bluefin tuna, which, if successful, would provide an inexpensive way to increase tuna production without removing juveniles from the wild.

Germ cell transfer is certainly an exciting technology, and one that may have a variety of uses in conservation biology. There are some drawbacks, however, to using germ cells for the purpose of de-extinction.

First, primordial germ cells are haploid; they either become sperm or eggs. When a sperm with an edited genome fertilizes an egg that does not have an edited genome (or vice versa), the offspring's diploid genome will have only one copy of the edited gene. The edits, therefore, may not be visible in the offspring's phenotype. To produce offspring with two copies of the edited gene, genomes from both the sperm *and* the egg have to be edited.

Second, the injected primordial germ cells are not the only primordial germ cells that make it to the sex organs. In the duck example above, the duck was the dad of the chicken, but his sperm were chimeric—some of his sperm were duck sperm and other sperm were chicken sperm. When his duck sperm fertilized a chicken egg, nothing happened. No hybrid "duckens" were born. But, when his chicken sperm—those of his sperm that were derived from the chicken primordial germ cells that were injected into his egg while he was a developing embryo—fertilized a chicken egg, a chicken was born. That chicken had a genome that was 100 percent chicken-like. But its father was, nonetheless, a duck.

Third, in the experiments that have been done thus far, scientists observed that the efficiency with which the injected primordial germ cells go on to become the next generation is poor. Only a small fraction of the eggs and sperm that are eventually made by the embryos develop from the injected primordial germ cells.

Mike McGrew of the Roslin Institute has a plan to overcome these obstacles. He is genetically engineering chickens that cannot make primordial germ cells. The only way these chickens would make eggs or sperm would be if primordial germ cells were injected during the appropriate developmental stage. In this way, he can produce hens in which 100 percent of eggs contain the edited genome, and cockerels in which 100 percent of sperm contained the edited genome. Mating these together would result in offspring that are 100 percent genetically engineered.

While there has been some success in transferring primordial germ cells between distantly related bird species, I imagine that there are still limits to how far this can be taken. Chickens, for example, may struggle to (and probably should not be caused to) lay eggs that contain developing moa or elephant bird embryos, for example. And there is little doubt that the hormonal and genetic environment within the mother—even for just the first twenty-four hours of development—plays some role in early embryonic development. This technology is exciting, however, and will certainly find use in the preservation of avian biodiversity, at the very least among chicken breeds.

And perhaps someday a chicken will be persuaded to lay an egg that contains a baby dodo. If that were to happen, the next question might be, just what is that chicken going to do with a baby dodo?

CHAPTER 9

MAKE MORE OF THEM

In 2004, a group of twelve distinguished scholars—conservation biologists, paleoecologists, mammologists, and community ecologists among them—met at Ted Turner's Ladder Ranch in the Chihuahuan Desert of New Mexico and developed a visionary plan for North American biodiversity. They proposed to reintroduce a small number of large-bodied animals, many of them endangered, into what little wild habitat remained on the continent. In doing so, they would protect North American biodiversity from continued decline. As a bonus, some endangered species would be provided a new, safe place to live and a better shot at survival.

Their premise was simple: big animals are integral to any ecosystem. Big animals play key roles in recycling nutrients, distributing seeds, turning over soils, and knocking down trees. Big animals are, however, missing from the North American landscape, largely due to terrible things that humans have done. To restore North America to a more balanced state, it is therefore necessary to restore big animals.

The group of scholars pointed out that restoration efforts tend to focus on reestablishing the flora and fauna that were present in North America when Europeans first arrived several hundred years ago. By that time, however, most of the big animals that had dominated the landscape throughout the Pleistocene ice ages were already gone. The group proposed looking further back in time to what they believed was a more appropriate baseline for North American restoration. A better target, they insisted, would be the Late Pleistocene—before human arrival and before the megafaunal mass extinction. The Late Pleistocene, they argued, was a time during which a diverse community of large herbivores maintained a diverse community of vegetation and were preyed upon by a diverse community of large carnivores. Naturally, the continent looked very different during the Late Pleistocene than it did when the first European colonists arrived.

Restoring North America to a Late Pleistocene baseline would be challenging, especially since many of the species that dominated the landscape at that time are now extinct. Not all of them are gone, of course. Some species survived, albeit in much more diminished ranges, for example, North American bison and giant desert tortoises. These species could be reintroduced wherever suitable habitat remains within their former range. Species that have gone extinct, such as camels, horses, and mammoths, could be replaced by proxies—living species capable of filling niches that were left vacant when the megafauna disappeared. Where reasonable proxies could be found, these species could be introduced into habitats that were once occupied by their extinct evolutionary cousins.

The plan for restoration was to start small and proceed in stages. First, Bolson tortoises (also known as Mexican giant tortoises) would be reestablished across the Chihuahuan Desert, which stretches from central Mexico northward through western Texas and parts of New Mexico and Arizona. The Bolson tortoise is North America's largest living terrestrial reptile. Although it was distributed across the Chihuahuan Desert during the Pleistocene, the Bolson tortoise is now restricted to a tiny, semiprotected refuge in north-central Mexico. Fortunately, the former range of the Bolson tortoise still includes some ideal habitat for reintroduction. Big Bend National Park in Texas, for example, used to be home to Bolson tortoises, and reintroduced tortoises could presumably get right back to the business of grazing on bunch grasses and digging burrows. It is unlikely that tortoise reintroduction would significantly alter the existing ecosystem of Big Bend National Park, other than to disturb the soil in a useful way. And it is unlikely that the tortoises would require much human intervention to survive. The most visible effect of Bolson tortoise reintroduction would probably be an increase in tourism to the park, as people realize they might be able to spot an eighty-year-old giant tortoise in its native habitat.

After the tortoise, the group planned to introduce horses, donkeys, and camels across the wilderness regions of western North America. Not just feral domestic horses and donkeys, but also their wild Eurasian cousins: the Przewalski horse and Asiatic wild ass. The group would also introduce camels—wild camels if possible, but domestic camels would suffice.

Why these species? When the ancestors of present-day horses and camels were living in North America (both horses and camels evolved in North America), woody plants were heavily grazed by large herbivores. This opened up space in which other types of plants could flourish, increasing floral biodiversity. A greater diversity of plants could sustain a greater diversity of herbivores, both large and small. And these, in turn, could support a greater diversity of carnivores. Large herbivores also act as efficient distributers of both nutrients and seeds. Their feet turn over the soil as they roam and run, their bodies transport seeds over long distances, and their excrement fertilizes the soil. Thanks in part to these animals and the roles they played within the ecosystem, Pleistocene North America was a mosaic of plant biodiversity that was capable of supporting a mosaic of animal biodiversity. Reestablishing horses and camels may help to restore this biodiversity.

Of course, the group was aware that introducing horses and camels to wild land in North America would be somewhat more controversial than introducing Bolson tortoises to desert ranches and US national parks would be. Feral horses and donkeys are considered by some people to be pests that compete with livestock. Any plan for reintroduction would have to balance the needs of the people who use the land with the potential benefits to the ecosystem. Strategies would need to be developed both to educate the public about why having these animals around might be good for the ecosystem and to teach people how to interact with these animals when they come into contact or conflict. Equally importantly, legal guidelines would be required to manage introduced populations and mitigate any potential negative consequences of reintroduction. At least some of the introduced species would *not* be native to North America—Bactrian camels, for example. Developing these strategies might therefore require new and creative thinking by legal scholars and wildlife managers. And finally, while Bolson tortoises probably won't need human intervention to maintain reasonable population sizes, populations of horses, donkeys, and camels could explode if left unchecked, with potentially devastating consequences to the ecosystem that their introduction was meant to preserve. After all, during their Pleistocene heyday, large herbivores were kept in check by large carnivores that are now extinct.

Which brings us to the next stage of the plan: cheetahs and lions.

And elephants.

African cheetahs, African lions, and Asian and African elephants. In North America.

Just as Bactrian camels were proposed as proxies for the extinct North American camel, *Camelops*, African cheetahs would take the place of the extinct American cheetah, *Miracinonyx*, and African lions would fill in for the extinct North American lion, *Panthera leo atrox*. Asian and African elephants would fill the niche once occupied by mammoths, mastodons, and gomphotheres.

To be clear, the plan was *not* to take animals from Africa or Asia and bring them to North America—this was one of the many angry accusations that came in the wake of the plan's release to the general public—but to identify and translocate animals already in captivity in North America to more realistically natural settings.

Needless to say, the plan to rewild North America did not pass quietly under the radar. Josh Donlan, who was the lead author on the two-page article[1] that appeared in the journal *Nature*, received the bulk of the public backlash. Donlan reported a pretty even mix between lovers and haters of the plan, with responses falling mostly within the range of what was predictable. There were, however, some surprises. Among the lovers were a handful of ranchers who were thrilled that they might be able to use elephants to keep the brush on their land at bay, as elephants would be much less expensive to operate than the heavy machinery they rely on at present. These ranchers were understandably less keen on the big cats.

FACILITATED EVOLUTION

The motivations behind the rewilding movement are similar to those that underlie my interest in de-extinction. Proponents of rewilding aim to restore biodiversity to ecosystems that have been negatively affected by extinctions. They hope that rewilding, by reestablishing lost biodiversity and re-creating missing interspecies interactions, will allow a much richer, more productive, and more diverse community of plants and animals to prosper. De-extinction could do the same thing, but with one small but important difference. The plan proposed by Donlan and his colleagues to rewild North America included the introduction of Asian or African elephants. However, Asian and African elephants never lived in North America and may not be particularly well adapted to the North American climate, which is much cooler than that in which they evolved. De-extinction also aims to introduce elephants into habitats in which present-day Asian and African elephants may not survive. But de-extinction will first prepare these elephants to live in a cooler climate by resurrecting adaptations that evolved in their cold-adapted cousins—mammoths—and inserting these adaptations into the elephants' genomes.

It is precisely in this way—by resurrecting adaptations from the past within the genomes of living organisms—that I imagine de-extinction as a powerful new tool both for biodiversity conservation and for the management of wild and semiwild habitats.

Take mammoth de-extinction, for example. Some advocates for mammoth de-extinction probably don't care what ecological role unextinct mammoths might play on the Siberian tundra. Some probably don't even care if they ever make it to the Siberian tundra, as long as they make it to a zoo or a park where they can be observed and possibly ridden. I, however, and others including George Church and Sergey Zimov, care very much about how unextinct mammoths—or, more correctly, genetically engineered Asian elephants—might change the Siberian tundra. In fact, their potential to invigorate the Siberian tundra is precisely why we are motivated to support this project.

So what would the Siberian tundra ecosystem gain from the introduction of cold-tolerant elephants? Working within the boundaries of his Pleistocene Park over the past few years, Sergey Zimov has shown how large herbivores—bison, muskox, horses, and several species of deer—can transform a mostly barren tundra into a rich grassland over the course of only a few seasons (plate 16). It's simple. They trample and graze the tundra, turning over the soil, dispersing seeds, and recycling nutrients. Their increased grazing stimulates the growth of grasses, which increases the density and nutrient quality of the available forage. Not all of the grass that grows can be consumed during the summer, leaving sufficient resources to support the animals during the Siberian winter. After the snow falls, the herbivores return regularly to the richest areas of grassland, trampling down the snow and eating everything beneath. Above ground, the grasses are consumed entirely. Below ground, the roots remain intact. In essence, Zimov's research has shown that the interaction between herbivores and arctic grasslands is self-sustaining. When one part of that interaction disappears, so does the other.

Zimov believes that the Siberian tundra could be transformed into rich grasslands reminiscent of the Pleistocene steppe tundra simply by returning large herbivores to the ecosystem. Revived steppe tundra would provide resources and habitat for other endangered species, including wild horses, saiga antelopes, and Siberian tigers. Zimov argues, however, that the missing critical piece to his Pleistocene puzzle is elephant-sized. Large herbivores play different ecological roles within a community than do smaller herbivores. Large herbivores knock down trees and trample bushes, for example, and transport seeds and nutrients over much longer distances than small herbivores can.

There is another, potentially more significant benefit to having large herbivores graze the Siberian tundra. Although the uppermost layers of the Siberian soil freeze and thaw with the seasons, the soil beneath these layers remains relatively constant in temperature throughout the year. This constant temperature is roughly equal to the mean annual air temperature, with an important caveat. During the winter, ambient air temperatures in Siberia can be as low as −50°C; however, snow sitting on top of the permafrost insulates the permafrost soils from this bitter cold, keeping them warmer than they would otherwise be during this time of year. Prior to the extinction of mammoths and other ice age megafauna, this snow would have been completely removed in some places and trampled in others, destroying its heat-insulating properties. The soil temperatures would have been dramatically colder than they are today. Although the number of grazing herbivores in Pleistocene Park is too small to have this same effect, it is nonetheless clear at smaller scales: Zimov estimates that the soil beneath grazed land in his park is somewhere between 15° and 20°C colder during the winter months than that beneath ungrazed land.

Scientists estimate that there may be as much as 1,400 gigatons of carbon currently trapped in the frozen arctic soil—almost twice the amount of carbon that is in Earth's atmosphere today. As global temperatures rise, the permafrost is melting and the carbon trapped within that permafrost is being released. If Zimov is right, then reintroducing mammoths into Siberia—or rather, introducing cold-tolerant Asian elephants into Siberia—will actually slow the accumulation of greenhouse gases in Earth's atmosphere and therefore the rate of global warming.

Importantly, the scenario above does not require the resurrection of a mammoth. All it requires is a mammoth proxy: an elephant that's genetically engineered to survive in Siberia.

ONE PLUS MORE MAKES A POPULATION

One elephant will not convert a denuded landscape into a flourishing and diverse ecosystem, regardless of how many genes were altered and how well-adapted the resulting animal is to living in that environment. However, this is exactly what we will have when the first phase of de-extinction—creating a living organism—is complete: one magnificent, healthy, genetically engineered elephant. Getting this far was certainly no walk in Pleistocene Park. Now we have to do it again.

To forge ahead with the second phase of de-extinction—releasing populations into the wild—we need to answer three questions. First, how many individuals will be required to establish a healthy population of our resurrected species? Second, how genetically diverse will the population need to be in order for it to be sustainable? Finally, where and how will this population be raised and nurtured so that it can eventually be released into the wild?

Several options are available to create a viable population of genetically engineered individuals. In the absence of significant improvements in the efficiency of genome editing, it is likely that only one cell will end up with all the requisite genomic changes. We could make more than one animal using this cell by growing that cell into a colony of identical cells—this is often called a *cell line*—and then using multiple cells from that cell line to create clones via nuclear transfer. One drawback to this approach is that all of the animals born will be genetically identical and, consequently, our population will have no genetic diversity. As another option, we could breed the engineered individuals with individuals that are not engineered. This would have the benefit of increasing the population's genetic diversity but may result in the loss of the genetic changes that we worked so hard to engineer as nonengineered genomes are bred into the population. A third option would be to start from scratch and reengineer the genome edits into cells isolated from a different individual. This would also increase genetic diversity but might not result in an organism with the same or even the desired phenotype. Because every genome is different, and all of the genes within a genome interact with each other, edits that have the desired phenotypic result in one cell may not have the same result when interacting with the genome of a different cell.

Given how hard it will be to create even one genetically engineered individual, and given that it will be just as hard to create a second edited individual that is not a genetic clone of the first, perhaps we should take a step back and ask whether genetic diversity is actually necessary for a population to survive. Do we really need to worry about creating a genetically diverse population?

The answer is *probably*.

Genetic differences between individuals are the substrate for adaptive evolution. If everyone in a population has the same genotype, then everyone will also have the same, or an extremely similar, phenotype. Everyone will be equally likely to survive and to reproduce. Of course, everyone will also all be equally *unlikely* to survive. If a disease sweeps through the population, for example, everyone will be equally susceptible to that disease. If the environment suddenly changes—perhaps there is a severe drought and an important source of food disappears—no individual will be better able to adapt to that change in resource availability than any other individual will be. Populations with high genetic diversity are buffered against disease and environmental fluctuations. Some individuals in these populations will be more likely than others to survive and reproduce. The genetically diverse population will adapt and survive.

Are high levels of genetic diversity absolutely necessary, however? Low levels of genetic diversity have been linked to poor health, decreased reproductive success, and even physical abnormalities, such as the crooked tail that was frequently observed among Florida panthers prior to their hybridization with panthers from Texas. Some species, however, have extremely low levels of genetic diversity but little measurable consequence to their ability to survive. Polar bears, for example, have extremely low levels of genetic diversity, but they have had the same tiny amount of diversity for at least the past 100,000 years. During that time, polar bears survived two ice ages and the present warm interglacial period. Nonetheless, as the habitat to which they are very specifically adapted disappears, their lack of genetic diversity may be their downfall: the more genetically diverse a population is, the greater the chance that this diversity will recombine in new ways, resulting in phenotypes that can adapt to surviving in a different environment.

Clearly, genetic diversity and the adaptive potential that genetic diversity provides are important, and a healthy population cannot be made up entirely of genetic clones. While it won't be the simplest solution, the most likely solution to the diversity problem will be to engineer cells taken from different individuals and use multiple cells to create a genetically diverse population. When editing the genomes within these cells, we will need to be certain that the edits are made in both chromosomes of every cell that is used. That way, the population will be genetically identical at these *particular* loci, and the target phenotype will be maintained even after the population is released into the wild.

While genetic diversity will be important to consider when creating our resurrected population, we must also keep in mind that diversity is not the only factor that determines whether a species is sustainable over the long term. If we were to survey genetic diversity among living primates and use this information to decide which primate species is most in need of protection, the result would shock most of us. The primate with the least amount of genetic diversity is . . . *us*. Humans have almost no genetic diversity, whereas other primates, including chimpanzees and gorillas, are doing just fine. Engineering genetically diverse populations will be important for de-extinction, but, ultimately, it will not be as important as finding a stable, healthy, and sufficiently large wild space into which our population can be released.

FROM THE BIRTH OF ONE TO THE REARING OF MANY

The second phase of de-extinction involves not only creating multiple individuals, but also rearing and nurturing these individuals, moving them out of captivity, and establishing populations in the wild. Ideally, this second phase would culminate with the establishment in the wild of multiple genetically robust, healthy, self-sustaining populations that are resilient in the face of environmental change. Phase two is certainly not going to be easier than phase one.

As a start, the babies must develop into adults. They must develop both physically and behaviorally, taking on the characteristics that they have been engineered to express. Most likely, several generations will be born and raised in captivity before a sufficient number of individuals are available to be released into the wild. Populations living in captivity, possibly for decades, need not only to survive, but they must also learn how to live. The individuals that make up these populations need to learn how to feed and protect themselves, how to interact with others, how to avoid predation, how to choose a mate, and how to provide parental care to their own offspring. Understanding how a species might fare in captivity is therefore an important consideration in deciding whether a species is a good candidate for de-extinction.

Humans have lots of experience raising and breeding organisms in captivity. For decades, we have raised animals in zoos, farms, breeding centers, and even our own houses. This experience has taught us that species differ in how they respond to captivity. Some species thrive—they are healthier, live longer, and have more offspring than wild-living individuals of the same species. Other species suffer terribly—they have shorter life expectancy, rarely reproduce, and even develop psychological disorders such as the repetitive swaying or pacing that is often observed in polar bears in zoos. Understanding how our candidate species for de-extinction is likely to fare in captivity will be critical to the success of our project.

When rearing genetically engineered animals, it will be important to keep in mind that some of the traits observed in these animals—both physical and behavioral traits—may be a consequence *not* of their edited genomes but of the pressures of life in captivity. One fascinating example of how different selection pressures in captivity can change the way an animal *looks* involves a wild population of Russian silver foxes. In 1959, Dmitry Belyaev, a Russian biologist who would later become director of the Institute of Cytology and Genetics, Russian Academy of Sciences, took 130 wild silver foxes and began breeding them on a farm near his institute in Novosibirsk. With each generation, he allowed only the foxes that appeared to be the tamest to breed. After only four generations, foxes began to wag their tails as their keepers approached. Over the course of just a few decades, his population of wild silver foxes was transformed into a population of animals that whined, wagged their tails, and jumped into the arms of and licked the hands of their keepers. The behavioral transformation was fascinating, but so was their physical transformation. The younger generations included animals with floppy ears, rolled or shortened tails, and coat coloration that is not seen in the wild but is reminiscent of other *domestic* animals.

Animals that are born and raised in captivity tend to look and act differently than their wild cousins. Physical differences appear in animals that are bred in captivity, including having shorter intestinal tracts and brains that are different sizes from those of wild-bred individuals. Traits that advertise sexuality also tend to be less noticeable among captive-bred individuals, which can influence their ability to find and compete for a mate in the wild. The behavioral differences between captive-bred and wild-bred individuals may be even more troubling. Animals in captivity don't need to learn how to avoid predation, for example, and the absence of social conflict and the unnatural social structures that form in captivity can lead to changes in defensive or sexual behaviors. Without sufficient space or stimulation, some species become overwhelmed by stress in captivity and develop acute psychological disorders, such as wall licking by captive giraffes and self-mutilation by captive big cats and bears. Stress also affects the animals' physiology, often reducing their fertility or stopping reproduction entirely.

Captive breeding also leads to unintended genetic changes, which may complicate the interpretation of genome-editing results. Without the need to find food, avoid predators, or fight off disease, selection pressures on animals living in captivity are relaxed. Instead of favoring traits for survival in the wild, captive breeding favors traits that increase reproductive success in captivity. This is not ideal when the goal is to release these populations into the wild.

Given the problems that so many species experience in captivity, it would be useful to have some way to predict a priori how a species might fare during this stage of de-extinction. It is tempting to guess how an extinct species might fare in captivity based on how well their living relatives fare. However, data from zoos and other captive-breeding facilities show that even close evolutionary relatives can differ considerably in their response to captivity. Among cetaceans, for example, captive Fraser's dolphins and Dall's porpoises damage their bodies by throwing themselves against the sides of enclosure pools and refuse to eat, while bottlenose dolphins and finless porpoises seem happy and playful in captivity, with reproductive rates that are sometimes higher than those observed in wild populations.

Those species that do fare well in captivity, from an animal welfare perspective at least, tend to be those that flourish in close proximity to humans. They include species that are sometimes characterized as "invasive," such as rats and mice, and species that do well in urban environments. They are species that deal well with disturbance and are flexible in their reaction to predators and new resources. In a sense, species that do well in captivity are those that are not particularly likely to be extinct in the first place.

YET ANOTHER SET OF MAMMOTH CHALLENGES

Elephants, unfortunately, are among those species that do not fare well in captivity. Both African and Asian elephants live longer in the wild than they do in zoos. Elephants in zoos are prone to obesity, arthritis, and infections, particularly in their feet. Even worse, both living species of elephant struggle to reproduce in zoos. Their ovulation cycles become abnormal and unpredictable, and they have low fertility rates and high infant mortality rates. Many elephants living in zoos also show signs of psychological distress, including repetitive swaying behavior, hyperaggression toward other elephants, and a propensity to kill their infants. These animals are provided with food, water, and medical care, and yet it seems clear that their most basic needs are not being met.

Elephants are known to be intelligent, social, wide-ranging animals and have needs that are very difficult to satisfy within the confines of most enclosures. There is little reason to suspect that the physiological and psychological needs of elephants whose genomes contain some small fraction of mammoth DNA will differ considerably from those of elephants whose genomes have not been edited. If elephants are going to be used in future de-extinction projects, sincere efforts will be necessary to improve the well-being of elephants in captivity and elephants released into the wild. This includes both careful consideration in the design of any enclosure in which these animals will breed and live, and the establishment of sufficiently large numbers of these animals in the wild to satisfy them socially and intellectually.

The challenges of captive breeding are likely to vary considerably among species. For example, species that migrate annually over long distances may be particularly unsuited to captive breeding, as sufficient space for this behavior will be exceedingly hard to replicate in a captive setting. If migratory paths are learned by interaction with a social group, how should scientists replicate the process by which these behaviors are learned?

Passenger pigeons were not migratory birds. However, they did fly long distances in order to find forests with sufficient numbers of fruiting trees to sustain their large flocks. Fledgling passenger pigeons learned this behavior by opportunistically joining flocks as they passed overhead. In his presentation at the TEDx event in March 2013, Ben Novak presented a plan to teach resurrected passenger pigeons how to find food. He proposed painting hundreds or thousands of homing pigeons so that they looked like passenger pigeons and training these painted homing pigeons to fly over the breeding colonies. These "surrogate flocks," as he called them, would attract the attention of the fledglings, who would follow their instincts and join the flocks. The surrogate flocks would ferry the young passenger pigeons between feeding sites that Ben intended to set up across the northeastern United States. As the passenger pigeon population grew, Ben would gradually use fewer and fewer surrogate birds in his flock, until eventually only passenger pigeons remained, complete with behaviors taught to them by Ben via his trained, painted flock of homing pigeons.

The combination of captivity-induced stress, reproductive problems, genetic consequences of different selection pressures, and lack of appropriate social interactions in captivity perhaps explains why captive-breeding programs for conservation—those that aim to rear endangered species in captivity and eventually release them into the wild—have been so variable in their success. Some strategies to resurrect extinct behavioral traits in captivity, such as that proposed by Ben Novak to train passenger pigeons to find food, are so far-fetched that they might actually work. There is no doubt, however, that captive breeding will be another tall hurdle for de-extinction.

Yet, captive breeding may not be as high a hurdle as the step that would come next: releasing these genetically modified organisms into the wild and allowing them to fend for themselves.

CHAPTER 10

SET THEM FREE

The California condor was once found as far north as British Columbia, as far south as Mexico, and as far east as New York. The large-bodied bird fed on the remains of even larger-bodied ice age animals, including mammoths and horses. When these animals began their gradual decline toward extinction, so did the California condor. Eventually, the California condor was restricted to California, where it survived by scavenging the remains of large marine mammals, including whales and seals. Elsewhere across its former range, it disappeared.

As human populations boomed along the California coast during the nineteenth and twentieth centuries, the California condor did not fare well. A program to preserve their habitat was established in the 1930s, but it had little impact on declining condor populations, and by 1982, the total population of California condors had reached a startling low of twenty-two individuals. In a last-ditch attempt to save the condors from extinction, a partnership was formed between the US Fish and Wildlife Service, the Los Angeles Zoo, and the San Diego Wild Animal Park. This partnership established a captive-breeding program for California condors. The program began with several eggs and chicks taken from wild nests and a single wild adult. A few years later, a controversial decision was made to relocate all remaining California condors from the wild into the breeding program. The hope was to conserve as much genetic diversity as possible while that diversity was still around.

California condors have a slow reproductive cycle compared with other birds. They breed for the first time between ages six and eight, after which a breeding pair will produce one fertilized egg every year or two. The program attempted to increase the reproductive output of the captive population by implementing a trick known as "double-clutching." Female condors can be duped into laying a second and sometimes a third egg if the first eggs are removed from the nest. When the first egg was laid, breeders would move it to an incubator so that another egg could take its place in the nest.

Double-clutching worked for the condors, in that many breeding females did in fact lay more than one egg. As this first round of eggs hatched in the incubators, however, a new problem surfaced. Who would raise the hatchlings? Who would teach them how to be California condors? Some of the incubator-hatched eggs were placed with foster condor parents and this worked out well. However, there were too few potential foster condors to place each hatchling with a condor parent. The breeders would have to rear some of the baby condors themselves.

Rearing by humans is tricky, as too much close contact with humans during these early life stages can lead to imprinting—an unhealthy trust of humans on the part of the of the baby chicks. Chicks that are too trusting would be at a disadvantage after release into the wild. Humans can be pretty nasty, after all.

So, the human breeders became puppeteers. They watched videos of real condor parents interacting with and feeding their young. They learned and adopted an appropriately strict, condor-like parenting style, which they implemented as best they could using puppets that resembled the heads of adult condors. Chicks fed by puppets also participated in a mentoring and socialization program, in which they spent time in an aviary with condor-raised chicks and other adult condors.

The first captive-bred California condors were released into the wild of southern California approximately five years after all wild condors were taken into captivity. By the end of 2010, the number of California condors had increased from twenty-two to around four hundred, approximately half of which were living in the wild.

AND . . . RELEASE

By many measures, the California condor captive-breeding program has been a success. More condors are living freely in California today than would be alive anywhere had the breeding program not been established. The path to success has, however, been circuitous and expensive, and condors are by no means in the clear with regard to extinction risk. Many of the problems encountered by the condor program are directly applicable to de-extinction and are worth considering here.

First, given the challenges that the condor program experienced while trying to build the condor population, are there some species for which captive breeding will simply be too slow to be successful? And, for those species that do have slow reproductive rates, is there any way to speed up the process? The developmental interval between the first and second generations of elephants is very long. Sergey Zimov told me that this is what he finds most troubling about the mammoth cloning and engineering projects. I had a chance to talk with Zimov recently at a conference, just after he delivered an impassioned speech about how mammoths would transform his Pleistocene Park. Keeping his voice low, he admitted to me that he would actually prefer woolly rhinos to mammoths. Pointing to his long gray beard, he conceded with visible dismay that an animal capable of reproducing at age five (like a woolly rhino) rather than at age fifteen (like a mammoth) was more within what he considered to be his personal time-frame. Mammoths, he said, would be for his children to introduce to the park.

Obviously, de-extinction projects will proceed more quickly with species that reproduce frequently and have many babies at once. In the first meeting to develop the passenger pigeon de-extinction project, one of the traits that was suggested as a first target of genetic engineering was the number of eggs laid per mother per year. Passenger pigeons laid a single egg once a year. It was proposed that we try to double this, and make each bird lay two eggs at a time. Two eggs a year would certainly facilitate the early stages of a passenger pigeon–rearing project, as it would mean more animals to work with while the population was small. But, if one egg at a time led to billions of passenger pigeons, I'm not certain that we want to genetically engineer them to reproduce at twice their normal rate. Perhaps a simpler solution would be to practice double-clutching during the early stages of captive breeding, as they did in the condor project. After the passenger pigeon population grew to a reasonable size, we could simply revert to leaving that first egg in the nest to be reared by its parent.

Rearing the young is another challenge that was highlighted by the California condor project. What steps will need to be taken within a de-extinction project to identify or manufacture appropriate social surrogates? How important is rearing within a social group, and what will the effect be if that social group is not entirely natural? Will it be possible for the juveniles to avoid imprinting either to humans or the surrogate species? This is a particularly difficult list of questions, and the answers are likely to vary considerably from species to species. One way to minimize these problems may be to select for de-extinction only species that lack significant parental care, where behavior seems more likely to be genetically hard-wired than it is to be learned. That would be bad news for highly social elephants and not particularly good news for passenger pigeons, which bred in large colonies with up to 100 nests in a single tree and both parents tending to each chick. It does not, however, exclude all species as candidates for de-extinction. Most turtles and tortoises appear to have very little parental care, which has made them target species for a variety of captive-breeding and reintroduction programs. Intriguingly, releases of captive-bred sea turtles have been conducted for decades and have thus far resulted in zero successful establishments or reestablishments of sea turtle–breeding colonies. We simply do not yet understand the complexities of how some behaviors are learned.

Another issue raised by the condor breeding program concerns the extent to which breeding in captivity alters behavior. California condors raised by puppets, despite their time served in aviary-based mentoring programs, displayed different behavior toward humans than did their siblings that had the fortune of being reared by actual condors. Puppet-reared condors integrated poorly with the rest of the condor community. Rather than shy away from humans, they preferred to play with garbage, hang about on roofs chewing on loose tiles, and stare disdainfully at rock climbers from above. Of course, with California condors, altered behavior was measured by comparison with the behavior of wild-reared animals. What constitutes "natural" behavior for species that have never been observed in the wild?

Individuals that are raised by surrogate parents or by social groups comprising closely related species can also develop *species confusion*, where offspring develop behaviors that are more similar to behaviors of the foster-parents' species than to their own species. If the foster parent is simply an unengineered version of the same species, it may be extremely challenging to establish and maintain behavioral differences using genome editing, and it might become necessary either to manufacture a surrogate social group or to use a more distantly related species. After release, another significant challenge will be to enforce reproductive isolation between engineered and unengineered populations of the same species. If the natural ranges of these groups overlap, uncontrolled breeding may quickly erode any genetic distinction between them.

Another issue raised by the California condor program concerns the number of individuals that will need to be released in order to establish an effective, natural social structure in the wild. Now that there are more than 230 California condors alive in the wild, it is apparent that condors are social animals whose social structure is key to their survival. California condors mate for life, are highly protective of their mates and territories, and have a strong system of social dominance that determines who eats and when. This social structure did not become apparent to scientists until their population size was sufficiently large to allow it to form. Elephants are also strongly social animals. Females live and raise the young in large, multigenerational family groups. Within these groups, a dominant, older, and wiser matriarch is responsible for decision making, including where to go to find food and water and when to flee from a potential threat. If captive-bred populations are to survive in the wild, the initial release will have to include a sufficiently large number of individuals and a sufficient range in age and experience to allow these social structures to emerge.

The *Allee effect* is a biological phenomenon that sometimes acts on very small populations. If a population is subject to an Allee effect, it is only stable when it is larger than a certain threshold size. If the number of individuals dwindles below that size, the population plummets suddenly to extinction. The idea behind the Allee effect is that individuals are more fit when the population is big. When the population is small, individuals are more susceptible to predation, have a harder time finding a mate, and are less efficient at discovering sources of food.

The extinction of the passenger pigeon in eastern North America is often cited as an example of the Allee effect in action. As hunting pressure increased and passenger pigeon populations declined, individuals may have been easier targets for predators like hawks without the protection of their enormous flocks. Also, deforestation around the turn of the century meant that food in the form of fruiting beech and oak trees was becoming increasingly difficult to find. The smaller populations of passenger pigeons may have been less capable of locating this limiting resource than larger populations would have been. If it is true that passenger pigeons suffered as a consequence of the Allee effect and are only capable of surviving in very large populations, it could be tough to generate a sufficiently large population in captivity for passenger pigeons to ever establish self-sustaining populations in the wild.

Ultimately, the goal of de-extinction is to create populations that are able to survive in the wild without human intervention. The California condor provides more insight here. Since their reestablishment in the wild, California condors continue to require considerable veterinary care. A main contributor to their decline, at least in the latter half of the twentieth century and continuing to today, has been lead poisoning. The birds eat fragments of lead bullets left behind in carcasses and gut piles, and the lead builds up in their system, making them very sick and eventually killing them. Lead bullets are being phased out of use in California, and the ban should be fully in effect by 2019. For now, however, lead bullets remain in use. Every condor is removed from the wild twice each year and put through extensive veterinary testing. Many of these birds are returned to captivity for treatment, specifically for the removal of lead from their blood. Without this expensive and time-consuming treatment, the birds would die.

The sad fact is that most species reintroductions fail. *Why* reintroductions fail is likely to vary from case to case. If whatever it was that drove the species to extinction or near-extinction in the first place was not accurately identified or, as in the case of the condor, has not been adequately resolved, then the reintroduction has little chance of succeeding. Genetic, behavioral, and social anomalies may arise as a consequence of captive breeding, and these may make captive-bred individuals unfit for life in the wild.

GENETICALLY MODIFIED ORGANISMS AS ENDANGERED SPECIES

One additional and important consideration in this final phase of de-extinction is how resurrected organisms (or organisms with resurrected traits) will be regulated once they are ready for release. Most countries have laws that regulate the release of nonnative species within their borders. These regulations will almost certainly apply to de-extinction projects that involve the genetic engineering of existing species. Elephants whose genomes code for mammoth traits would probably be regulated as invasive species in Siberia, for example. Bucardos, however, might not be considered invasive should they be released into their former range. Instead, the practicalities of bucardo re-release might be determined by public land use laws and perhaps even endangered species laws. There is another option, however. Because the organisms will all have been genetically engineered to some extent, they will fall into the category of *genetically modified organisms,* or GMOs, and—perhaps—under the purview of GMO legislation.

The world has widely varying opinions about GMOs, including whether they should be considered safe, how they should be managed, and which laws should apply to their use and distribution. This diversity of opinion is reflected in the scope of laws in place to regulate GMOs. The United States is among the least stringent countries when it comes to regulating GMOs and is one of the largest producers of GMO crops. The European Union has some of the most stringent GMO regulations in the world, but countries vary considerably within the EU with regard to whether they believe those regulations are necessary or even fair. New Zealand has among the strictest regulations of GMOs. If moa are someday brought back to life, would these rules exclude them from being released into the wild? Or from being eaten by New Zealanders?

To explore the regulatory labyrinth to be navigated by resurrected organisms, let's consider the release of passenger pigeons in the northeastern United States. If we succeed in creating band-tailed pigeons whose genomes contain some passenger pigeon DNA—for simplicity, let's refer to them as passenger pigeons—they will have been produced using genetic engineering technologies, specifically genome editing and cloning via primordial germ cell transfer. They will, from a scientific perspective, be genetically modified organisms. However, they may also be nonnative, they will almost certainly have some effect on the environment into which they are released, and they may even be considered endangered. To which regulatory agency can we turn to determine whether and where we can release them?

First, let's consider their status as GMOs. In the United States, a framework was developed in the 1980s to regulate GMOs using existing federal agencies. These agencies were charged with evaluating the safety of and risk associated with specific types of GMOs. Following this framework, GMO-derived food and medicines are regulated by the Food and Drug Administration, while organisms that are characterized as GMO pesticides—for example, plants that have been engineered to express genes that make themselves resistant to viruses—fall within the remit of the Environmental Protection Agency. The risk that these GMOs pose to the environment and to agriculture is evaluated by the Department of Agriculture.

While this framework seems straightforward, one important limitation stands out: these laws apply only to GMOs that are intended to be consumed. So unless our motivation for passenger pigeon de-extinction was to farm passenger pigeons and sell them to the hungry masses, resurrected passenger pigeons would not be considered GMOs in the United States, at least by federal laws.

For the purpose of exploration, let's say we do want to bring back passenger pigeons so that we can sell them as food. In this case, our edible passenger pigeons would be reviewed by the FDA as GMOs. If the FDA review found that GM passenger pigeons contained some sort of unusual substance that was not found in band-tailed pigeons (the unmodified food product), the FDA could then establish mechanisms to oversee their farming and production. FDA laws would *not,* however, address what might happen should passenger pigeons escape from the farm and take up residence elsewhere.

At present, GMO regulations in most countries apply only to food. If, after escaping their FDA-regulated farms in the northeastern United States, our grown-for-food passenger pigeons flew across the Canadian border to reestablish free-living populations across their once-native range, whether they were "allowed" to enter Canada would be determined by whether they were defined as an invasive species, not by their status as GMOs.

Our intent is, of course, not to farm and sell passenger pigeon meat, but to establish natural populations of passenger pigeons in the wild. While this intent excludes resurrected passenger pigeons from federal regulation as GMOs, it does not exclude them from local GMO regulation. In the United States, local GMO regulations can be much more inclusive in their definition of GMOs, and many local laws ban GMOs even when they are not meant to be used as food. In Marin County, California, for example, an ordinance is in place that bans all GMOs except those that are used in enclosed, licensed, medical facilities. The ordinance defines GMOs as "an organism or the offspring of an organism, the DNA of which has been altered or amended through genetic engineering." That definition would seem to exclude resurrected passenger pigeons from moving to Marin, even though this is where Stewart Brand and Ryan Phelan live, and where Revive & Restore—the organization that is behind their de-extinction—is based.

In most countries, resurrected species will be regulated under environmental statutes—invasive species, public land use, and endangered species laws—rather than under GMO laws. It is relatively easy to understand why the first two categories apply: most (but not all) resurrected species will be nonnative, and many release programs will include public land. However, is it appropriate for resurrected species (or species with resurrected traits) to be protected as endangered species?

Endangered species protection for resurrected species sounds good, but it would likely be a double-edged sword. Increased regulation would make it harder for breeding facilities and wildlife managers to manipulate the species, even if such manipulation were designed to benefit their recovery. At the same time, protection would provide a variety of benefits. For example, in the United States, it is illegal to kill individuals belonging to a protected species without an explicit permit. In addition, other federal agencies have to explicitly consider how their decisions or regulations would impact protected species. Critical habitat for the species also has to be identified and protected, and officially sanctioned plans have to be developed to recover their populations.

For the passenger pigeon to receive protection under the US Endangered Species Act, it would first have to be listed as an endangered species by the US Fish and Wildlife Service. Species qualify for listing if they are affected by one or more of five factors: (1) real or imminent habitat loss; (2) overexploitation; (3) disease or predation; (4) inadequate protection by other regulatory mechanisms; and (5) other natural or manmade attributes that will affect its continued existence. Without a good idea of what its range might be, it is hard to know whether the passenger pigeon would qualify for listing on the grounds of having insufficient habitat. Given that regulations governing GMOs are not meant to protect the status of the GMOs (but instead to protect the world from GMOs), it would almost certainly qualify on the grounds of factor 4: insufficient protection. Passenger pigeons would probably lack genetic diversity, which would be a factor (manmade, in this instance) that could affect its continued existence (factor 5). Also, back-breeding with band-tailed pigeons could lead to the re-extinction of the passenger pigeon, and therefore band-tailed pigeons might be considered natural factors that also could affect their continued existence (factor 5).

So, the passenger pigeon would probably qualify for listing within the United States as *endangered*, but what about as a *species*? This is trickier. Does a band-tailed pigeon with a bit of passenger pigeon DNA inserted into its genome constitute a separate *species*? The Endangered Species Act does not wade into the murky waters of defining a biological species but instead considers any subspecies or even (for vertebrates) "distinct population segments" as separate species, for the purpose of listing. Because resurrected passenger pigeons would really be band-tailed pigeons with a few or more extinct genes thrown in, this—as a distinct population segment of band-tailed pigeons—is the most likely avenue by which they might qualify for listing.

In the unfeeling eyes of the law at least, passenger pigeons probably would qualify for protection as endangered species in the United States. Does protection by this mechanism make sense, however, in light of the purpose of the legislation? The Endangered Species Act and similar regulations were intended to protect living endangered species. Just as many food and drug laws were not developed with GMOs in mind, endangered species legislation was not developed with de-extinction in mind. Forcing existing regulations to absorb unextinct species, with their myriad additional challenges and uncertainties, could cause these often precariously balanced sets of rules and regulations to come tumbling down, with potentially dire consequences to the existing legal structure and to those species that are currently under protection.

Clearly, regulations to protect endangered species were not envisioned to protect man-made species. But is a species that contains resurrected traits truly man-made? It might have an altered genome sequence, but the alterations evolved, naturally, within the genomes of now-extinct species. The traits themselves are natural, but the genetic combination of these traits and the genome of a living species is man-made. This semantic limitation—the necessity of distinguishing completely between natural and unnatural—is a limitation of existing environmental laws that illustrates how unprepared the regulatory sphere is for de-extinction.

The International Union for the Conservation of Nature currently lists the band-tailed pigeon as a "species of least concern." This is good news for the first phase of passenger pigeon de-extinction, as it limits what regulations govern the use of band-tailed pigeons in genetic-engineering and captive-breeding programs. But it's less good news for the last phase of de-extinction. If the band-tailed pigeon were itself endangered, then the Endangered Species Act would have a convenient mechanism to provide the passenger pigeon with the benefits of protection but without the bureaucracy. To provide some flexibility for captive-breeding programs of endangered species, the Endangered Species Acts allows experimental populations of endangered species to be considered "nonessential," in that the survival of that particular population is not absolutely necessary for the survival of the species. Nonessential populations must live in a geographic area that is completely separate from the essential part of the species range. Conveniently, living separately from other populations of band-tailed pigeons will also be important for ensuring the survival of passenger pigeons genes within band-tailed pigeon genomes.

To summarize, is not at all clear how endangered species regulations will apply to resurrected species or traits. De-extinction certainly does not fit neatly within any existing regulatory mechanism, and different types of de-extinctions (cloned bucardos versus slightly modified band-tailed pigeons) are likely to fall into different regulatory categories and to require new interpretations of existing rules. It is also unlikely that there will be widespread agreement among or even within countries about what can and should be done to regulate de-extinction and manage resurrected species. Only one thing is certain: genetic modification of living things is possible, and genetically modified organisms for the purpose of conservation will soon exist.

There is some good news for resurrected mammoths. If mammoths are brought back and introduced into a private park, whether that park is in the United States or in northeastern Siberia, these mammoths would not be regulated either as GMOs or by national environmental laws. Visitors to the park may even be allowed to hunt and eat the resurrected mammoths without breaking any national laws. Local laws might apply, so the location of the park could be important. For now, however, Sergey Zimov's plan to rewild his Pleistocene Park in Siberia with genetically modified elephants faces no obvious regulatory obstacles.

TOWARD REWILDING AND ECOLOGICAL RESURRECTION

The idea to rewild North America with living species that would act as proxies for the extinct, native megafauna of the Late Pleistocene made a big splash when it was first introduced in 2005. Reactions varied from overwhelming enthusiasm to almost violent rejection. After a few months, rewilding gradually disappeared from the headlines of the mainstream media and became relegated to specialist, scientific reports. Some of these were continuations of the ongoing debate about whether or not rewilding was a practical tool for the purposes of conserving biodiversity, or about what the target baseline for rewilding projects should be. (Should we aim for a Late Pleistocene-like landscape, or a pre-European-like landscape?) Other reports contained success stories, such as the removal of invasive species and reestablishment of native species on islands that were sufficiently small for such projects to be tractable. While the scale of these successes was much smaller than that envisioned by Josh Donlan and his colleagues in their 2005 article, these successes were nonetheless important. They demonstrated that rewilding—and, by extension, de-extinction—is a strategy that can change landscapes in dramatic and fundamental ways.

Of course, the ecological changes brought about by the release of resurrected species into wild habitats might not always be those that were envisioned at the start of the de-extinction project. When a resurrected species (or a species with resurrected traits) is introduced into an ecosystem, its introduction will change that ecosystem, just as its extinction did. However, the ecosystem will have evolved since its disappearance, and how the ecosystem will respond to its reappearance is not entirely predictable. Given the knowledge that we cannot completely control the results of our experiments, should we proceed? When is the risk of de-extinction worth the potential reward?

CHAPTER 11

SHOULD WE?

On March 15, 2013, a TEDx event was held at the National Geographic Society headquarters in Washington, DC, to celebrate and inaugurate the idea of de-extinction.[1] The event coincided with publication of Carl Zimmer's *National Geographic Magazine* cover story, "Bringing Them Back to Life."[2]

On March 16, 2013, "de-extinction" hit the headlines like only new wars, missing airplanes, or resurrected mammoths can. Those of us who were involved with the event anticipated that this might happen. Our biggest concern was to try to limit hyperbole so that our message could be heard by anyone who cared to hear it. Those among us who supported de-extinction—and not all of us in the program did—hoped that de-extinction would become a tool that the conservation community could add to their existing arsenal of defense mechanisms against contemporary extinctions. We worried, however, that we would instead be seen by this community as, at best, jostling to compete with them for already limited resources and, at worst, providing a convenient excuse for the rest of the world to care even less about protecting endangered species.

At a rehearsal on the day before the big event, Ryan Phelan and Stewart Brand, who organized the event, passed around a media package that contained concise and clear (and consistent) answers to what they predicted would be the most common questions. The night before the event, they and we (the speakers) hosted a select group of local and national media, politicians, and heads of conservation-oriented NGOs at an invitation-only kickoff event. With this, we hoped to persuade stakeholders that the science we would be presenting was real, that we cared deeply about and understood the historical and political context in which we were operating, and that we were very much aware of how our message might affect—both positively and negatively—conservation movements within the United States and internationally. We wanted to be clear that our intention was not to sensationalize science fiction but to engage stakeholders and the public in a reasonable, scientifically validated debate.

The TEDx event was brilliantly organized, academically interesting, and a lot of fun. My talk was not particularly positive about the prospects of bringing exact replicas of extinct species back to life. Other talks were more enthusiastic, predicting major advances in no time at all. Mike Archer, an Australian scientist who is leading a group called the Lazarus Project, presented brand new results that were released to the media in Australia *during* his presentation. Mike's research team had just succeeded in creating embryos from frozen cells of the extinct Lazarus frog, which was an awesomely peculiar amphibian that swallowed its tadpoles and later regurgitated fully metamorphosed juvenile frogs. Although the Lazarus frog embryos did not live for more than a few days, Mike insisted, correctly, that this was a major step forward in Lazarus frog de-extinction. Ben Novak unabashedly displayed his obsession with the passenger pigeon by presenting a detailed plan for exactly how he was going to release them into the wild after they were brought back to life. Distinguished professors of conservation biology, philosophy, law, and ethics brought up orthogonal points relating to whether or not de-extinction was realistic, dangerous, or (and?) morally reprehensible.

The initial reaction to the TEDx event was mostly one of unadulterated excitement. The mammoth was going to be cloned! (Clearly, nobody paid attention to my presentation.) The passenger pigeon was going to darken the skies once more! (Everyone seemed to have forgotten Michael McGrew's talk, during which he explained to the audience that birds cannot be cloned.) The world was going to be saved! (Perhaps no one remembered Stanley Temple's talk, which highlighted the need for careful consideration of the ecological consequences of introducing an extinct species to an ecosystem that had continued to evolve in its absence.) George Church was going to change the world! (Yes, that's probably true.)

Doom and impending catastrophe sell more newspapers, magazines, and documentaries than do happy visions of the future. This is not new to conservation scientists. Irreversible climate change, imminent extinctions, the disappearance of the forests and the animals in them—those are the topics of headlines. Solutions, success stories, reintroductions—these stories are relegated to the space beneath the classifieds. The headlines about de-extinction were clear. De-extinction is *dangerous*. Some scientists say it's a bad idea. It could and probably will go horribly wrong, just like in *Jurassic Park*. And given that it is definitely happening and the mammoth is being cloned and passenger pigeons are about to darken the skies again and Lazarus frogs will soon be barfing up their babies, the public should be *scared*. The public should stop de-extinction from happening! At the very least, the public should make it known that they are aware of the sneaky, dangerous stuff going on in those ivory towers and that it doesn't make them happy.

I started to get hate mail. I was both terrified and surprised by this. My presentation had been one of the most negative, pointing out, as I have in this book, all of the challenges facing those who wish to bring extinct species back to life. In my many media interviews that followed, I did my best to maintain a positive yet skeptical outlook. The media, with some exceptions,[3] have often been less than appreciative of my skepticism. I've been in several interviews in which the interviewer spent a lot of energy trying to get me to say something sensational or controversial—as if "Yes, I am working with others to bring back something similar to the mammoth and passenger pigeon" is not sensational enough.

I also got fan mail. Several people wrote to congratulate us on our bravery and foresight. People offered to send us bones, teeth, and feathers that they'd found while gardening. Students wrote heartfelt letters begging to join the lab so that they could be involved with bringing back the passenger pigeon. Mike Sweeny, who is the executive director of Nature Conservancy in California, contacted me to see whether there was some way that his organization could help with de-extinction in California. There was a tremendous, positive outpouring of support.

The hate mail was equally sincere. I was accused of playing God. I learned that I was going to bring about the end of the world. I was informed that I was a menace to society and should be stripped of my academic credentials. One note even suggested that I should be the first meal for any saber-toothed cat that we managed to bring back.

Professional scientists don't send hate mail, but they do publish hate papers. Several very smart, very highly regarded scientists came out in force against the de-extinction movement. Professor Paul Ehrlich is an eminent scientist at Stanford University and the president of Stanford's Center for Conservation Biology. He is perhaps best known for his ominous predictions about what will happen to the world if human populations continue to grow as they are today. Ehrlich emphatically refused an invitation to attend a workshop at his own university that was sponsored and organized by Professor Hank Greely, an equally eminent law professor who specializes in biotechnology law. In fact, Ehrlich recommended so forcibly that no one in his department make the slightest pretense toward supporting the idea of de-extinction that not a single Stanford biologist was present at the meeting, despite the fact that it took place on their doorstep *and* aimed to address precisely the topics that infuriated Ehrlich.

Months later, Ehrlich agreed to a public debate of sorts with Stewart Brand, who, it turns out, had studied as an undergraduate under Ehrlich when Ehrlich first joined the Stanford faculty in 1959, and whom Ehrlich still considers to be a good friend. The debate was not a conversation but, instead, a pair of written essays that presented opposing views about whether de-extinction should go forward.

When I first read Professor Ehrlich's essay, I was surprised by many of the issues he chose to highlight. What surprised me was that many of the problems he noted, while certainly valid, important, and deserving of consideration, are not unique to de-extinction. The problems he highlighted are the same problems that come up every time a new tool for biodiversity conservation is proposed, an embattled realm in which Ehrlich himself is no stranger. While Ehrlich begins his essay by taking issue with the financial cost of de-extinction, his more powerful objections are to the indirect but potentially more insidious costs—the potential costs to society, to endangered species, and to endangered ecosystems.

From the meanest to the most profound, all objections to de-extinction arise from genuine fears and deserve to be addressed. Below, I attempt to speak to those concerns that I hear most often or that I feel are the most central to the ongoing debate. Not every question can be answered, and this is one of the most troubling truths about de-extinction. Certainly, there will be costs associated with de-extinction, including costs that we have yet to imagine. I feel strongly, however, that there is one very important cost that Professor Ehrlich and the often-anonymous writers of hate mail fail to cite: the cost of doing nothing.

WE MIGHT REVIVE DANGEROUS PATHOGENS

Since we cannot know exactly why the last few individuals of an extinct species died, is there a chance that it was a dangerous pathogen that killed them? And if we bring these individuals back to life, might we also bring back that dangerous pathogen?

Probably not. In addressing this, it is important to consider where pathogens would be preserved. Most pathogens do not integrate into the genomes of the organisms they infect. Instead, they attack a specific part of that organism—the lungs, for example, or liver, or blood cells. If tissues could be revived from an extinct organism and those tissues happened to contain a pathogen, then it is possible that the pathogen could also be revived. One of the recent mammoth finds in Siberia contains what looks like blood, which contains what looks like blood cells. If this organism had been infected by a blood-borne pathogen, then this blood-like substance could have included what looked like cells from blood-borne pathogens (it did not, as far as I know). However, it has not yet been possible to revive cells from extinct species, as the genetic material within them is simply too degraded. This will apply to pathogen genomes as well. No recovered pathogen cell will be sufficiently intact that it is capable of coming back to life.

Genomes do contain some viruses that integrate into the genome. Our own genomes are full of such viruses, the vast majority of which are not harmful. If we were to extract DNA from a bone and sequence everything that was recovered, the mixed pool of extracted DNA would include DNA from the animal whose bone it was, DNA from infectious pathogens that were present at the time of death, and DNA from anything that got into the bone—including other pathogens—during burial and excavation. All of this DNA, however, at least all of it that is ancient, will be fragmentary and damaged, as expected for ancient DNA. Any ancient viruses or pathogens that were preserved within that sample would certainly not be in any state to be infectious.

DE-EXTINCTION IS NOT FAIR TO THE ANIMALS

This might be true. Animal welfare does need to be considered explicitly when developing a plan for de-extinction. In previous chapters, I outlined some of the ways that animals might be exploited or harmed in the course of this work. Some species—such as Steller's sea cow—may be terrible candidates for de-extinction, simply because it would be impossible to resurrect them without causing unnecessary animal suffering. As technology advances, this may become less true. For example, technology that allows in vitro rather than in vivo gestation would eliminate the requirement for cross-species gestation. From an animal-welfare perspective, the captive-breeding stage is likely to be one of the most challenging steps of de-extinction. Better understanding of the basic needs of animals in captivity and of how we can minimize the effects of being raised in captivity once animals are released in the wild will be key to the success of de-extinction. These are areas of active research, and advances will come. As of today, the possibility that too many animals might suffer remains a serious obstacle to de-extinction.

WE SHOULD PRIORITIZE CONSERVATION OF SPECIES THAT ARE ALIVE TODAY

In 2014, I participated in a conference in Oxford, UK, on the importance of megafauna—both extinct and extant—in maintaining the ecosystems in which they live. The keynote speaker was George Monbiot, a journalist and environmental activist who writes a weekly column for the *Guardian*. Monbiot's speech was a lively and impassioned plea to support rewilding in Europe. At an emotional apex and with tears forming in his eyes (at least in my memory), he roared angrily, "Those billionaires that are funding de-extinction—they should instead be investing their millions to introduce the Asian elephant to Europe!"

I agree with him about the elephants. One of Monbiot's most salient points was that the European vegetation had evolved in conjunction with a type of elephant—a mammoth—and, since elephants are missing at present, we should put them back if we can find a place for them. I agree. If elephants, whose native habitat is dwindling, could be introduced into pockets of Europe where efforts are already under way to rewild, why not do so? Asian elephants might even survive without genetic manipulation in parts of Europe.

But the billionaires? Who and where are they? And may I have their number? As of this moment, I know of no de-extinction project that is being funded, much less being funded by billionaires. The biotech development that is going on in the Church lab is only possible because the technology itself has another purpose—specifically the purpose of curing human disease. Money for my group to sequence genomes from passenger pigeons and band-tailed pigeons has been cobbled together out of my small research budget from the University of California, some funds from private foundations that are dedicated to developing techniques for the assembly of ancient genomes, a donation of several thousand dollars from Revive & Restore, and voluntary time from people like Ben Novak, Ed Green, and others working in the group. The bucardo project has some support from a local hunting federation, but certainly not enough to fund an entire de-extinction project. If billionaires are investing in de-extinction, I haven't heard about it. But I would like to hear more.

Should de-extinction compete for resources aimed at the preservation of living species and habitats? Absolutely not. Is de-extinction competing for resources with these organizations? Today, the answer is very clearly no. In 2014, the US government budgeted just under $414 million for all of its international conservation initiatives and exactly $0 for de-extinction research. Conservation International reports spending around $140 million every year, $0 of which is spent on de-extinction projects. The World Wildlife Fund spends around $225 million on its various international programs, none of which involve or are related to de-extinction.

The costs of later stages of de-extinction, including captive breeding and release and the long-term management of free-living populations, will be harder to tuck neatly away into the budgets of other projects. It is doubtful that breeding mammoths will lead to a cure for human genetic disease, for example, which makes it hard to justify mammoth-breeding expenses on a grant from the National Institutes of Health. When it's time to breed mammoths, new sources of money will have to be found. These sources are likely to be different from those that fund existing conservation initiatives. People give to causes they care about, and different people care about different things. The people who care about the plight of polar bears or pandas will probably not be the same as the people who want to bring passenger pigeons back to life. Hopefully, as de-extinction as we perceive it gains momentum, this will lead to the discovery of new sources of funding for conservation initiatives and a strengthened focus on the creation and preservation of wild habitats.

While the idea that de-extinction may spark interest in conservation—more precisely, in funding conservation research—is attractive, it also highlights an important weakness in the present strategy to fund de-extinction research. Today, this research is being performed by scientists on species that the scientists find interesting. However, private individuals are being asked to fund the work. Just as the lion's share (pun intended) of conservation funding goes to the most charismatic of endangered species, species will probably be selected for de-extinction based on their public appeal. People are likely to be far more interested in dodos and Steller's sea cows than they are in extinct kangaroo rats and land snails, although kangaroo rats and land snails are arguably more critical to the stability of their ecosystems than either dodos and Steller's sea cows were to theirs. Ultimately, our partiality toward charismatic megafauna will lead to a taxonomic imbalance among de-extinction projects that is not unlike the imbalance that exists in conservation work.

If de-extinction is to become a genuine weapon to be used in the war against contemporary extinctions, all sectors of society—and not just scientists—will need to work together to identify the resources to make it happen.

UNEXTINCT SPECIES HAVE NOWHERE TO GO

Unfortunately, many species that are candidates for de-extinction have no habitat in which to live. The more people there are in the world, the less space there is for other species. Deforestation and illegal hunting are significant problems in many parts of the world. If these are the problems that led to extinction in the first place, these problems must be resolved before that extinction can be reversed.

Some species require more space than it might be possible to find. Gray wolf populations are booming in Yellowstone National Park, where they are protected from humans. The park provides nearly 9,000 square kilometers of space for wolves, but *this is not enough*. As the wolves jostle for territory and dominance, they expand beyond the park's borders. When they get out, they cause mayhem and get shot. When dominant wolves get shot, it upsets the pack structure and dynamics. Gray wolves cannot quite come to a sustainable equilibrium in a space the size of Yellowstone.

Finding sufficient amounts of suitable habitat will certainly present a challenge for some de-extinction projects. This should not, however, preclude further assessment of the suitability of other species for de-extinction. Nor should it deter efforts to improve habitat by removing invasive species or enforcing anti-poaching or anti-deforestation laws. On the contrary, highlighting this problem in the context of de-extinction may act as a beacon for new investment and new solutions, which would also benefit existing conservation projects.

RELEASING UNEXTINCT SPECIES WILL DESTROY EXISTING ECOSYSTEMS

To this concern, I respond with an emphatic "maybe." Certainly, a thorough assessment of the environmental impact of releasing new species into the wild should be done before a de-extinction project begins. Assuming the candidate species for de-extinction is an animal, the assessment should include analyses of what and how much it is likely to eat, with which other species it will compete for resources, where and when it will sleep, by what means and how far it will move, what will eat it and what are the consequences of it being eaten, whether it will act as a vector for disease, and what effect it will have on nutrient cycling, pollination, the microbial community, and so forth. Regardless of how thorough and careful these assessments are, there will be unanticipated interactions between species and unanticipated consequences to the ecosystem. This is unavoidable. When the species went extinct, the ecosystem in which it was once a part evolved to accommodate its absence. Other species, sometimes even invasive species, moved in. Reintroducing the extinct species may upset the existing dynamics within that ecosystem, but to claim that that ecosystem will be "destroyed" might be going too far. Yes, species introductions change ecosystems—that is often the point of the introduction. To this end, a risk assessment will not ask whether an ecosystem would change (it would), but how it would change, how other species would be affected, and whether the reintroduced species would be sustainable within that ecosystem.

Completing such an assessment is likely to reveal that some species are poor candidates for de-extinction. Some species would be too destructive to fit within the confines of today's people-dominated world; imagine sixteen-foot-tall short-faced bears wandering around downtown Los Angeles. Some species simply have nowhere to go; the Yangtze River dolphin cannot be placed back into its natural habitat unless there is dramatic improvement in the water quality of the Yangtze River. Some species may require more long-term investment than it is possible to secure. We may find that so little is known about the behavior and ecology of some species that the risks of environmental catastrophe far outweigh the benefits of their return to the ecosystem.

If a reintroduction does have catastrophic consequences, we could simply remove that species from the ecosystem using whatever means necessary. *Re*-extinction is certainly an extreme tactic, but it calls on expertise that we already know we have. Of course, it may not be that simple. Once an organism is released, it will start to affect the ecosystem into which it has been introduced. It's doubtful that everyone will agree about whether these changes are good, or even acceptable. Society as a whole will have to decide whether removal is necessary, and this will not be an easy decision to make.

Consider the example of beavers in Great Britain. Until recently, beavers were extinct in Great Britain. Beavers were driven to extinction in Great Britain some 400 years ago by humans, who valued beaver fur and medicinal glands and loathed beavers. Beavers are destructive; they cut down trees and use these to build dams, which cause rivers and streams to flood. Dead beavers were better beavers, at least to sixteenth-century Britons, and so beavers disappeared. Then, in 2006, beavers were discovered living along the River Tay in Scotland. In early 2014, a family of wild beavers was spotted in Devon, in the southwest of England, playing in the River Otter. It is believed that both of these beaver populations were established after deliberate and illegal release from private collections.

Along the Rivers Tay and Otter, residents differ widely in how they feel about the beavers. Some residents are quick to identify the positive impacts they've seen on the environment since the beavers reappeared. They point out that, by building dams along the rivers, the beavers have created new habitat for frogs that lay their eggs in the shallow, slow-moving ponds formed by the dams. These frogs and their eggs are, in turn, an important food source for insects, birds, and fish, which some residents claim have increased in abundance since the beavers' return. The beaver dams have also begun to reestablish local wetlands, which the residents hope may help to control flooding along the rivers. Other residents, however, dislike the beavers. These residents point out that beaver dams block migration routes used by salmon and trout, and may actually increase rather than mitigate flooding, with devastating consequences for riverside farms.

While beavers, fish, and agriculture coexisted in Britain for centuries, the British countryside has changed considerably over the last 400 years. Thanks to these changes, it is not at all clear that coexistence can resume.

So what to do? Should the illegally released beavers be removed from the British countryside, or should beavers be introduced to even more rivers? This has been a difficult question to answer. As a member of the European Union, Britain is under pressure to reintroduce native species that have been driven to extinction locally. The wild-living beavers may, in fact, already qualify for legal protection under EU laws. Within Britain, England, Scotland, and Wales get to decide for themselves what to do within their borders, and there is no consensus. Wales is considering allowing beavers to be introduced into the Welsh countryside, while the English government has established an official program to capture the beavers along the River Otter and remove them to captivity. Along the River Tay in Scotland, some 300 beavers now make their homes. The Scottish government is set to decide, soon, whether they get to stay.

The example above highlights yet another significant challenge that society will need to resolve if de-extinction is to move forward: when is it clear that a de-extinction experiment has failed? With beavers, the environmental impacts of release can be inferred from habitat in which the beavers still live. This will not be true when the organism to be released is completely extinct and, consequently, the risk that it all goes horribly wrong will be, admittedly, greater.

I want to circle back to the beginning at this point and restate something I said in the first pages of this book. While re-extinction is certainly an option, and one that quells the deepest fears of some de-extinction skeptics, I worry that people might resort to this drastic measure too quickly. Interactions between species may take years to develop. Ecosystems into which a resurrected species is introduced may become destabilized initially and only much later re-establish the interactions between species that were the goal of the de-extinction project. These experiments will take time, and I hope that we can be patient. It is natural, however, to fear what we do not know and cannot predict. Being patient will not be easy.

Concern about the appropriateness of hands-on environmental stewardship is not unique to de-extinction. Conservation strategies can be thought of as a continuum between entirely managed ecosystems (think "gardening") and allowing nature to fend for itself (think "preserving"). De-extinction is a disruptive strategy and, as such, requires some amount of gardening. However, like other disruptive strategies—including rewilding, managed relocation, and island restoration—de-extinction can play a role across nearly the entire continuum. Some species will require constant gardening, while others will require little to no intervention to be sustainable once established. Regardless, all disruptive strategies are inherently risky, as there is always a chance that a heavy human hand may do more harm than good. Purely preservationist strategies are, however, also risky. What if sufficient habitat can't be preserved? What if species do not re-establish populations in the habitat that is preserved? Few habitats have avoided completely the effects of human population growth, suggesting that, at some level, intervention has already occurred. Further intervention may be required simply to reduce the damage that has already been done.

Island restoration projects, such as two that are taking place off the coast of Mauritius, are proving that intervening can work. On Round Island and Ile aux Aigrettes, conservation biologists are working to remove invasive species and re-establish populations of native species. But there are problems. Native plants are slowly being replaced by invasive plants in the absence of the extinct giant tortoises that once thrived on the islands. The native plants grow slowly and close to the ground and have small tough leaves that are difficult for tortoises to eat. They also fruit when grasses—a main source of food for tortoises—are not abundant, which increases the likelihood that hungry tortoises would disperse their seeds. In the absence of giant tortoises, nonnative plants have outcompeted the tortoise-adapted native plants, many of which are now on the brink of extinction.

To restore the missing interactions between native plants and giant tortoises, the research teams introduced different species of giant tortoises that still survive in other parts of the Indian Ocean, hoping that these giant tortoises would functionally replace the extinct Mauritian giant tortoise. The introduced giant tortoises immediately took to their new habitat, preferring to graze on the non-native plants that lacked defenses against tortoise herbivory. They also ate the fruits of native species. Stands of ebony, which had struggled to survive in the absence of a large herbivore to disperse its seeds, have started to appear throughout the Islands.

IF DE-EXTINCTION IS POSSIBLE, THE RATE OF EXTINCTION WILL INCREASE

This moral hazard argument presents a horrible view of people. It assumes that that at the slightest (and I mean *slightest*) hint of a quick fix, no matter how not-so-quick and not-quite-a-fix it is, people will give up trying to preserve endangered species. Sure, legislation to protect endangered species is complicated, confusing, sometimes misguided, and too often out of date. But, it is hard to imagine that people who care about biodiversity conservation would suddenly stop doing so should de-extinction become possible.

Of course, there are many people who simply don't prioritize biodiversity conservation, and others who have some stake in seeing species removed from protection. In these cases, one might imagine how the idea of de-extinction might be manipulated to further a specific political agenda. The notion that politics or big business might use biotechnology to manipulate rules, regulations, and public sentiment is, of course, not unique to de-extinction.

WE ARE "PLAYING GOD"

As an epigram to the first edition of the *Whole Earth Catalog* in 1968, Stewart Brand wrote, "We are as gods and might as well get good at it." Like many of the ideas that motivate Stewart, this line, which he conceived while reading anthropologist Edmund Leach's book *A Runaway World*, was meant to make people imagine, with bold optimism, a future that was different, pleasant, and full of wonder. But, he did not want them to stop there.

Stewart wanted to motivate people to act, using that bold enthusiasm, to make real the future that they had just imagined.

Stewart's issue with science and society, then and now, is their deference to the status quo. Their detachment. His argument is simple and positive: we can make a better future, but not by standing by and waiting for it to happen. We—everyone—must participate. It is our responsibility to use our intelligence and our advanced technology for good.

The "playing God" argument is not one that that has emerged in response to de-extinction but, instead, is an argument that arises frequently in response to technology that is new or not well understood. This argument can be religious, but it is often a metaphorical accusation—"playing God" may simply mean using powerful tools without understanding the full implication of those tools.

In the specific case of de-extinction, the accusation of playing God concerns human manipulation of nature. By engineering new organisms, by altering the structure of biological communities, and by altering the course of today's extinction trajectory, we are messing with things that we simply don't understand and therefore probably shouldn't be messing with. Importantly, de-extinction does not mark the beginning of human manipulation of nature. With the earliest attempts at domestication of gray wolves in Europe some 30,000 years ago, our species began manipulating the genetics of other organisms to our advantage. Most of the food we eat has been genetically engineered—albeit by breeding and not by genome editing—to suit our tastes and to meet the growing demand for more. Species introductions, whether purposeful or accidental, have been happening since we first built boats and learned how to navigate from one place to another. And the extinction trajectory on which we are heading is, arguably, itself human induced.

I believe that what motivates this argument in the case of de-extinction is the fear of losing control. This is a reasonable concern. It is, however, a concern that should be expressed and addressed rationally, taking advantage of the scientific process.

THE PRODUCT OF DE-EXTINCTION WON'T BE THE SAME THING AS THE ORIGINAL SPECIES

That is correct. It won't be the same.

In Stewart Brand's half of his written debate with Paul Ehrlich, he writes: "If it looks like a passenger pigeon and flies like a one, is it the original bird?" My answer is no, it's not the same, and by this point in the book it should be clear why my answer is no. Crucially, however, *I don't care* that it's not the same thing as the original, and I'm pretty sure that Stewart doesn't care, either.

The task ahead is not to make perfect replicas of species that were once alive. First, it is technically not possible to do so and is unlikely ever to become technically possible to do so. Second, there is no compelling reason to make perfect replicas of extinct species. The goal of de-extinction is to restore or revive ecosystems, to reinstate interactions between species that no longer exist because one or more of those species are extinct. We don't need to create exact replicas of extinct species to achieve this goal. Instead, we can engineer species that are alive today so that they can act as proxies for extinct species. We can revive adaptations from the past—adaptations that arose by chance and were refined by evolution—in species that are still alive today.

In fact, there is no reason to restrict this technology to de-extinction. If living species are threatened by a lack of diversity or by an inability to adapt quickly enough to a rapidly changing climate, why not facilitate their adaptation as well?

The American chestnut tree is a great example of the power of genome engineering in conservation. Around the year 1900, the accidental importation of a fungus from Asia wiped out nearly every single American chestnut tree. The airborne fungus kills the tree by forming cankers in the bark that cut off the flow of nutrients from the ground. New shoots may grow from surviving roots, but none of these escapes the deadly fungus. Thanks to genetic engineering, American chestnuts are now on the verge of making a dramatic comeback into the eastern deciduous forests of North America. Led by Bill Powell and Charles Maynard of the State University of New York in Syracuse, the American Chestnut Research and Restoration Project has genetically engineered several new strains of American chestnut that are increasingly resistant to the fungus. In 2006, this team planted the first fungus-resistant chestnut seeds in the wild. Today, there are more than one thousand genetically modified American chestnut trees growing in the state of New York.

A BRAND OF POSITIVISM

Regardless of how feasible it really is, de-extinction has succeeded in forcing us—by "us," I am referring here to scientists who hope, as I do, that our research will have a positive environmental impact—out of our comfort zones, exactly as Stewart Brand envisioned it would. Stewart, of course, would like to see de-extinction do more than that. His goal is for de-extinction is that it will become "a reframing of possibilities as momentous as landing humans on the moon was."[4] Certainly, if it does become possible to resurrect extinct species or to coax living species to express extinct traits, our perception of what it means to be "extinct" will change fundamentally. The most momentous change, however, will be in our attitudes toward living species—this, I believe, is what Stewart is referring to when he speaks of possibilities reframed. Suddenly, we will have the technical know-how to engineer sustainability into threatened populations. Will improving rather than protecting species become the new objective of biodiversity conservation? If we turn to the past to identify traits that can be used to improve the plight of living species, where will we draw the line between preventing versus reversing extinction? And will we care?

This, I believe, is why people like me are so captivated by the idea of de-extinction. Not because it is a means to turn back the clock and somehow right our ancestors' wrongs, but because de-extinction uses awesome, exciting, cutting-edge technology to take a giant step forward. De-extinction is a process that allows us to actively create a future that is really better than today, not just one that is less bad than what we anticipate. It is not important that we cannot bring back a creature that is 100 percent mammoth or 100 percent passenger pigeon. What matters is that—today—we can tweak an elephant cell so that it expresses a mammoth gene. In a few years, those mammoth genes may be making proteins in living elephants, and the elephants made up of those cells might, as a consequence, no longer be isolated to pockets of declining habitat in tropical zones of the Old World. Instead, they will be free to wander the open spaces of Siberia, Alaska, and Northern Europe, restoring to these places all of the benefits of a large dynamic herbivore that have been missing for eight thousand years. De-extinction is a markedly different approach to planning for and coping with future environmental change than any other strategy that we, as a society, have devised. It will reframe our possibilities.

De-extinction will, of course, be risky. We don't know and cannot predict every outcome of resurrecting the past. The conservation success stories of the present day prove, however, that taking risks can be deeply rewarding. Removing every living California condor from the wild was an extraordinarily risky strategy to preserve the species, but one that undoubtedly saved them from extinction. Restoring gray wolf populations to Yellowstone National Park was both risky and, to a degree, unpopular, but the park is now flourishing in a way that it had not since its establishment in 1872, when wolves and other predators were actively exterminated. Allowing deer, cattle, and other wild animals to take over abandoned land in Europe was touted as both crazy and dangerous, but these reestablished wilderness areas stimulated a widespread shift in attitudes toward wildlife. They inspired new policies aimed at protecting natural spaces and the species that occupy these spaces. How will the world react when the first genetically engineered elephants are strolling casually through Pleistocene Park?

I can't wait to find out.

ACKNOWLEDGMENTS

When I signed on to write this book several years ago, my goal was simply to answer the question that I have been asked repeatedly since my first days working with ancient DNA: "Is it possible to clone a mammoth?" I could not have predicted that the idea of de-extinction would become so popular—and such a seemingly realistic goal—over the course of the book's creation. It has been both fascinating and exhilarating to be part of these early days of de-extinction research, both as a scientist and as a de-extinction storyteller. I am indebted to the many researchers and thought leaders who have been behind this wave of enthusiasm and, in particular, to Ryan Phelan and Stewart Brand of Revive & Restore, whose efforts to move de-extinction forward are unparalleled.

Actually writing the book has been both more challenging and more fun than I thought it would be. I appreciate all of those who read early chapter drafts and provided critical feedback. David States, Jacob Sherkow, Alberto Fernández-Arias, George Church, Tom Gilbert, Tony Ezzell, and Molan Goldstein all provided comments, corrections, and criticism that helped to make this book better.

I thank the team at Princeton University Press for their continuous encouragement and tireless enthusiasm for this book. Alison Kalett has been a wonderful editor throughout the process, a constant source of enthusiasm, and a pillar of support when necessary. Jessica Pellien has been a joy to work with, as have Katie Lewis, Quinn Fusting, Betsy Blumenthal, and the entire team. I am delighted to have had the opportunity to work with them all.

I am grateful to Tyler Kuhn and Love Dalén for allowing me to take advantage of their superlative photographic skills. I've taken thousands of photos while working in the field, but none of them come close to exposing the raw beauty of the arctic landscape as theirs do. Thanks also to Mathias Stiller, Alberto Fernández-Arias, André Elias Rodrigues Soares, and Sergey Zimov for sharing their images. Each of their images reveal critical aspects of the de-extinction process that words are inadequate to describe.

I am also indebted to the people in my lab at UC Santa Cruz, some of whom are mentioned in the preceding pages, for tolerating my partial absenteeism, in particular as the final deadlines approached. Many thanks, both for keeping the research going and for not spending all of our money while I wasn't paying as close attention as I should have been.

Finally, I am grateful to my large and extended family for their support and encouragement throughout this process. In particular, I would like to thank Ed Green, my partner both in life and in running our lab, for his encouragement, enthusiasm, and advice as the chapters came together, for the many days that he took over childcare so that I could spend those few extra hours in front of the computer, and, of course, for tolerating and even supporting the craziest of the de-extinction projects taking place in our lab. If it's a boy, I promise to name the first unextinct pigeon after you, Ed.

NOTES

PROLOGUE

1. Piers Anthony, *The Source of Magic (Xanth)* (New York: Ballantine Books, 1979).

2. "A new organization, Revive and Restore, formed by the Long Now Foundation with the help of the National Geographic Society and advised by a group of respected scientists, has been created to examine the potential for a new branch of zoology: de-extinction." *Times* (London), 8 March 2013, http://www.thetimes.co.uk/tto/opinion/columnists/benmacintyre/article3708288.ece.

CHAPTER 2: SELECT A SPECIES

1. Svante Pääbo, the director of the Max Planck Institute for Evolutionary Anthropology in Leipzig, Germany, and the leader of an international project to sequence the complete genome of a Neandertal, wrote an editorial in the *New York Times* in which he argues that as sentient beings, Neandertals have the same rights as humans and should not be cloned. His op-ed, "Neandertals Are People, Too," was published on 24 April 2014.

CHAPTER 7: RECONSTRUCT *PART OF* THE GENOME

1. This is one of the most fascinating and potentially the longest-lasting impact of the *Technology Review* report: ben-Aaron states that while elephants have fifty-six chromosomes, mammoths have fifty-eight. In fact, we have no idea how many chromosomes a mammoth has, and we are unlikely to know the answer until a high-quality version of their genome has been sequenced and assembled. Nonetheless, the "fact" that mammoths have fifty-eight chromosomes is widely reported on the Internet. I can only assume the source is ben-Aaron's article, as no reference or citation is ever provided.

CHAPTER 9: MAKE MORE OF THEM

1. This article, "Re-Wilding North America," appeared in the 18 August 2005 edition of *Nature.* Josh Donlan is listed as the article's only author, but a footnote points to a long list of notable conservation biologists including Harry Greene of Cornell University, Joel Berger of the Wildlife Conservation Society, Carl Bock and Jane Bock of the University of Colorado, David Burney of Fordham University, Jim Estes of UC Santa Cruz, Dave Foreman of the Re-wilding Institute, Paul Martin of the University of Arizona, Gary Roemer of New Mexico State University, Felisa Smith of the University of New Mexico, and Michael Soulé of the Wildlands Project.

CHAPTER 11: SHOULD WE?

1. TEDxDeExtinction was organized by Ryan Phelan and Stewart Brand of Revive & Restore. All of the presentations are available from the Technology, Education, and Development (TED) Web site and from http://tedxdeextinction.org/.

2. Carl's article was published in the April 2013 edition of *National Geographic Magazine*. In addition to his story, the feature includes some delightfully nostalgic photography by Robb Kendrick that depicts species that might be targets for de-extinction research.

3. I realize I am being overly harsh here. There have been several good articles written about de-extinction over the last several years. Carl Zimmer's work in *National Geographic Magazine*, which I mention above, is excellent. Also, Nathanial Rich wrote a thoughtful and nuanced piece about de-extinction in the 2 March 2014 edition of the *New York Times Magazine* that I think is among the best of what has been presented thus far about de-extinction.

4. This quote is from Stewart's "Point" in the "Point/Counterpoint" series in which he and Professor Paul Ehrlich agreed to participate. It was published on 13 January 2014 as part of Yale University's *Yale Environment 360*.

INDEX

Illustrations are indicated with **bold face**